FROM POVERTY TO PROGRESS

UNDERSTANDING HUMANITY'S GREATEST ACHIEVEMENT

Michael Magoon

Upward Press

Cataloging Publication Data

    Michael Magoon
    From Poverty to Progress / Michael Magoon

Library of Congress Control Number: 2021900034

EPUB Direct Sales 978-1-958206-00-3
EPUB Amazon 978-1-958206-01-0
Paperback 978-1-958206-02-7
Hardback 978-1-958206-03-4
Audio 978-1-958206-04-1

Credit and permissions are listed on the "Credits" page and are considered a continuation of the copyright page.

Published by Upward Press 2022, Boise, Idaho

E-book originally published in 2021

Book Series

This book is the first in an extended series of books about history, technological innovation, economic growth and progress. Each book in the series has a specific focus, but they all have a common perspective. This book explains my theory on the origins and spread of progress and overviews world history from that perspective.

After the first book, each of the subsequent books focuses on either government policies to promote progress or the history of progress in a specific time and region. "The Five Keys to Progress" is the central unifying concept of the series.

Published Books
From Poverty to Progress: Understanding Humanity's Greatest Achievement

Forthcoming Books on Policies to Promote Progress
*Upward Mobility: A Radical New Agenda to Uplift the Poor
    and Working Class*
    (publication: by early 2023)
Promoting Progress (publication: by early 2024)

Forthcoming Books on the History of Progress
Beginnings of Progress: The Rise of Commercial Cities (1200 – 1830)
Britain, Its Empire and the Industrialization of Progress
Progress Transforms America
Progress Transforms Europe
Progress Transforms Asia and the World

Website for Book Series

You can find additional content related to this book series at: **frompovertytoprogress.com**

With a **free** subscription to this website, you get:

- Large discounts on audiobooks and e-books
- Free book samples (E-book and audiobook)
- Access to Podcasts and Blog posts about related content
- Plus more.

Additional Content and Graphics

To keep the price of this book as affordable as possible, some content has been moved to the author's website. On this site, you can find:

- Full-color graphics
- Extra content and graphics
- Bibliography
- Recommended Reading List
- Access to summaries of related books
- Books that Influenced Author the Most

To access this content, scan the following QR code or go to **frompovertytoprogress.com** and search for "Additional Content"

Praise for This Book

"Progress is not a matter of optimism; it's an empirical fact of history, obscured by seeing the world through headlines rather than data. And it is not a natural process but the effect of distinctive technological and political circumstances. Magoon has made a valuable contribution in adding to our understanding of the facts and causes of the most important development in human history."

Steven Pinker, Johnstone Professor of Psychology, Harvard University, and the author of *The Better Angels of Our Nature* and *Enlightenment Now.*

"The concept of progress is perhaps the most important human idea. Michael Magoon's new book gives an excellent account of the origins of progress, its root causes, as well policies for how we can keep it going. It will change your thinking about progress and its relevance to your life."

Tyler Cowen, author of *Stubborn Attachments* and *The Great Stagnation;* Named by The Economist in 2011 as one of the top 36 most influential economists of the decade.

"In his book, Michael Magoon documents and explains the amazing progress of our time. Magoon's concept of the 'Five Keys to Progress' gives us a powerful new perspective on the historical causes of progress, how wealthy nations can keep it going and how developing nations can enjoy greater progress… A good read!"

Johan Norberg, author of *Progress* and *Open: The Story of Human Progress*, The Economist's Book of the Year, 2021

"In an Age of Despondency, this book's deep faith in progress is like a breath of fresh and hopeful air. May Magoon be right."

Joel Mokyr, author of *A Culture of Growth: The Origins of the Modern Economy* and *The Enlightened Economy: An Economic History of Britain 1700-1850*

"This book offers a radical challenge to the current—and very pessimistic—academic interpretation of history. It stresses the importance of technology, cooperation, and competition and the primacy of Western civilization in raising the living standards of the majority of humankind. All readers, whether they agree or not, will find it a refreshingly new perspective on history."

Daniel R. Headrick, author of *Technology: A World History*

"The world is a complicated place. But a handful of basic principles help make sense of it and reveal the road to progress. In this book Magoon does much to identify the wellsprings of material progress. I recommend it — especially to those who have begun to doubt the possibility of progress."

Jan de Vries, author of *First Modern Economy*

"Magoon's call for hope in the world should make us all appreciate what we have inherited from our shared human past. With that hope, we could share an even more promising future."

Peter Bellwood, author of *First Farmers* and *First Migrants*

"We tend to forget that technological change has vastly improved our standards of living over the course of history and dwell upon the losing end of innovation. Against this background, Magoon's rich narrative offers a compelling reminder of how societies have prospered by elevating the successful to models worthy of emulation.

Highly recommended."

Stelios Michalopoulos, author of *Ethnic Inequality*

"A fascinating addition to our understanding of the connection between long-term history and economic development. Michael Magoon forcefully reminds us of the progress made by humankind. Backed by a wealth of facts and data, he delves deep into the ingredients that enabled societies to achieve this.

Highly recommended."

Areendam Chanda, author of *Early Starts, Reversals and Catchup in the Process of Economic Development*

Contents

INTRODUCTION

Progress is humanity's greatest achievement. It has transformed our lives in so many positive ways, but we take it for granted, refuse to admit its existence, or even claim that it is bad. Far too many of us have become "progress deniers."

After reading the above paragraph, I know that some readers who are skeptical of progress will immediately want to move on to another book, but please give me five minutes of your time to see why you can benefit from reading this book.

Progress is real, and it matters to you.

We do not see all the progress that is around us, not because it is not there, but because we choose not to see it.

Part of the reason we cannot see the existence of progress is because we compare the problems of today to either an idea in our head about how life *should* be, or to nostalgic memories of how we think life *used to be*. But reality can never match the beautiful thoughts that the human brain can imagine. Nor can reality match the rosy nostalgia for how life supposedly used to be.

In general, among people who are skeptical of or hostile to progress, those on the Left and young people compare the problems of today with an idea in their heads about how life should be, while those on the Right and older people compare the problems of today to nostalgic but distorted memories of how they think life used to be.

Neither side can see the existence of progress because they are constantly bombarded with a distorted view of reality that reinforces their prior assumptions. The media, social media, political activists, interest groups and political candidates all have a strong self-interest in dwelling on the negative and sensational. An understanding of and appreciation for progress get in the way of their message and their self-interest. They are much better served by reinforcing their viewer's distorted view of reality for their own gain.

These institutions have learned that the best way to get attention is to magnify a problem until it appears to be an existential threat to us all. Then they follow up with a relentless drumbeat "proving" that the threat is becoming more common and more dangerous. They know that fear, anger, resentment and hate generate viewers, votes and money.

Many people, perhaps a majority, understand that progress exists. They know that their material lives are much better than in previous decades or generations, but they do not often think about it. They do not see the concept of progress as relevant to themselves. They take progress for granted and go on with their daily lives. They do not see that progress is under threat from the beliefs that I described above.

Worse, some believe that progress *does* exist, but that it is bad for society and threatens our future survival. Because of these views, they implement policies that try to protect us from progress. While they no doubt have good intentions, the policies these people implement do great harm by undermining the very foundations that help people to live better lives.

These views being broadcast to us are untrue, unhealthy and socially destructive, but they continue because they serve the self-interest of many individuals and institutions. For many of them, the concept of progress is an existential threat to their livelihood, so they fight any hint of it as strongly as possible.

Many people confuse this deliberately distorted view with an objective view of reality. All of these views are based upon what psychologists call "cognitive biases" towards pessimism and threat. People do not realize the enormous self-interest that many institutions in our society have in reinforcing these cognitive biases. So we react to what we perceive as reality, not to what reality actually is.

People with cognitive biases often avoid interaction with others who do not share their views. Rather than confronting these cognitive biases, many people segregate from others with healthier viewpoints both in the real world and in social media. Others spend endless hours alone watching television news or ideological rants. They often attack those who try to help them see a better world, calling them stupid, naïve, uncaring, or even immoral.

Rather than confronting the bias, they think that reality is the problem. The very things that they think make them feel better, actually make them feel worse. This causes far too many people to go into a downward spiral of unhappiness, hopelessness and disconnection from the real world.

The feedback loop caused by cognitive biases towards pessimism and threat, unrealistic views of what is possible, inaccurate memories of what life used to be like, isolation from alternative viewpoints, and institutional self-interest create a powerful psychological feedback loop. This feedback loop creates what can only be described as an alternative reality that seems very real but does not exist.

It bears a spooky resemblance to the film *The Matrix*. But whereas the Matrix in the movie created a wonderful life for people who were actually living in terrible circumstances, our real Matrix does the opposite. It takes people living in the best circumstances that humanity has ever lived in and creates the impression that "everything is going to hell." So, of course, they do not believe in progress. In the movie, the Matrix was a computer-generated dream world built by AI machines so they could keep people compliant while machines harvested our bio-energy. The real Matrix, however, has been created by millions of years of evolution that wired our brains to think in a certain way.

The fundamental problem is that the human brain did not evolve for happiness and appreciation for living in a world of abundance and progress. Nor did it evolve to solve problems in complex modern societies.

The human brain evolved to enable our Hunter-Gatherer ancestors to survive and reproduce on the African Savannah. This world was full of daily threats to survival.

Anyone who focused on appreciating their lives would soon get eaten by a predator, attacked by a stranger, or starve. Anyone who was relentlessly focused on threats to their survival was more likely to survive long enough to pass on their genes to the next generation. This way of thinking was very useful for our distant ancestors, but it causes serious cognitive biases in the modern world.

Humans have evolved with a brain that responds well to threats, and the media, social media, political activists, interest groups and political candidates have gotten really good at manipulating us with exaggerated threats. In fact, this has become their business model. They keep doing it because it works. This business model generates viewers, votes and money: critical resources that these individuals and institutions need to maintain their success.

Getting past these cognitive biases is not easy. An entire new sub-discipline of psychology called "Positive Psychology" has emerged to understand and treat similar conditions.

Many positive psychologists recommend Cognitive Behavioral Therapy (or CBT) as a way to overcome these cognitive biases. According to the official American Psychological Association (APA) website, there is strong evidence that CBT leads to "significant improvement in functioning and quality of life."

According to the APA, CBT consists of the patient:

- Learning to recognize one's distortions that are creating problems, and then to reevaluate them in the light of reality.
- Gaining a better understanding of the behavior and motivation of others
- Using problem-solving skills to cope with difficult situations
- Learning to develop a greater sense of confidence in one's own abilities.

CBT also has the benefit of treating a wide range of problems, not just one. It is useful for treating serious mental illness as well as the far more common "worried well." I would argue that those who are skeptical and opposed to progress are a substantial portion of the "worried well."

Unfortunately, implementing CBT on a mass scale is not practical. We need a better, more scalable option than CBT.

I believe that option can be found in the concept of progress and the "therapy" is the objective study of history so we can learn what caused that progress. We need to use a little bit of data plus a change of perspective to clear out the cognitive biases that make us unhappy, angry and resentful in a world full of abundance and progress.

Of course, I am not claiming that the study of progress actually is a therapy. The study of progress will obviously not cure anyone with a serious mental illness. Nor should the study of progress be seen as a replacement for face-to-face therapy by a trained professional. Nor am I claiming that anyone who is at all ambivalent about progress suffers from a mental illness. And nor has the APA endorsed it.

However, I am claiming that the skeptics and opponents of progress are suffering from many of the cognitive biases that can be compared to the mindset of people who benefit from CBT. I am claiming that the fundamental principles of CBT can be applied to more intellectual endeavors like the study of progress. Learning to "recognize one's distortions that are creating problems", "learn problem-solving skills" and "learn to develop a greater sense of confidence" are exactly what the Western world needs right now in abundance.

The study of progress is a functional equivalent of CBT. The study of progress is akin to self-therapy at scale for the "worried well."

To see the progress that surrounds us, we need to shift focus to study how our ancestors actually lived. Only by comparing today's life to the actual lives of previous generations can we fully appreciate the progress that we live in today.

Once we look at actual metrics comparing today's material circumstances to our ancestors', we can see that we live in a world of progress.

Coming to a deeper understanding of how our ancestors created progress and kept it going can also give us a sense of gratitude. Our ancestors worked really hard, and we are all beneficiaries of their efforts.

Positive psychologists, philosophers and religious thinkers have repeatedly stressed the importance of gratitude. Gratitude is a powerful emotion because it shifts our thoughts away from our own problems to something more important and more constructive. Ironically, the more one focuses on one's own problems, the more debilitating those problems seem. By shifting to something much bigger than oneself, one realizes that one's own problems are nowhere near as bad as they seem.

Even better, learning how our ancestors built progress gives us a toolkit for solving most of today's problems. We can see that in situations far worse than our own, our ancestors learned practical strategies for making their world a better place. Most of these strategies were very focused on solving short-term local problems, but as word got around, they had positive effects on the entire society.

When we clear out our cognitive biases, we can see that life in the past was actually pretty terrible. It was full of people with the same dread and worry that we have today. It was full of problems that were daily threats to survival that we do not have to deal with as much today.

When we clear out our cognitive biases, we learn that many people in the past who were animated by a strong desire to make the world a better place actually made the world worse. And far more of them failed to have any real impact at all.

They failed because they were so focused on how the world should be, that they did not seek to understand the world as it is. Instead, they tried to transform the world to match the beautiful idea that only existed between their ears. Utopian ideas do not make the world a better place; they almost always fail and many times they actually make the world much worse.

This does not mean that we should give up on our desire to make the world a better place. Far from it; we desperately need the passion of ideals to drive us forward. Only that passion can give us the energy to keep trying to solve big problems despite our repeated failure to do so.

It does mean, however, that we need to focus more on what has worked in the past, and then do more of it in the present and the future. It is only when we understand how much better things are now that we are open to a rational exploration of what worked in the past.

We need to combine the passion to make the world a better place with results-based experimentation at scale using methods that worked well in the past. This is neither optimism nor pessimism. It is merely focusing on what works.

When we clear out our cognitive biases, we can learn that, though very real problems still exist, those problems are actually fewer in number and milder in severity than the ones previous generations had to deal with. And previous generations did not have all the technologies, skills, organizations and scientific knowledge that we have today.

If we have an awareness of the progress that previous generations have passed down to us, a feeling of gratitude for benefitting from their efforts and a willingness to learn how they achieved that progress, we are in a much better position to solve the problems of today.

While today's problems seem insurmountable, we must never forget that previous generations solved far bigger problems with far fewer resources. Knowing this, we can look forward to the future with both hope and the necessary problem-solving attitude.

Looking to the past enables us to overcome our cognitive biases towards the present. The good old days were never really that good. Utopias cannot exist in the real world. The world as portrayed by the media and politics is highly distorted and dangerously so.

This book is not about psychology, the media, social media or politics. But in order to make my case, I knew that I first needed to clear away some of the false beliefs that you probably had before you started reading this book.

This book explains how humans invented progress, and how we can keep it going. It claims that, by understanding the past, we can copy what works to make the future better. By understanding the past, we can also clear out all the biases that undermine our self-confidence, and shift to solving problems rather than worrying about crises.

So now that that is done, let's get on to the real point.

Progress is Real

People living in Western nations today have a level of affluence far surpassing anything ever seen on planet Earth. Even the poor in Western nations have a level of affluence that is higher than that which all but the richest people had in 1970.

All across the world, nations are being transformed from oppressive poverty to a level of affluence that was once only possible in Western nations. Japan, South Korea, China, India, Singapore, Botswana, Chile and Puerto Rico all transformed themselves within one generation. Even in some of the poorest nations of Sub-Saharan Africa, levels of education, health, literacy, sanitation, longevity, transportation, communication, and housing are rapidly improving.

The Good Old Days Were Pretty Terrible

This progress is a startling transformation compared to how humans have lived over the past hundred thousand years. For most of human history, there was little if any progress.

Our ancestors were engaged in a constant struggle for survival. Most people lived on the edge of subsistence. The vast majority of our ancestors devoted the majority of their waking hours to acquiring enough food to eat. The quest for the next meal dominated their lives.

Most humans lived in societies that changed very little over the course of their lifetime. The vast majority of our ancestors were trapped in circumstances that were very similar to their parents' and their grandparents' lives. They knew that their children and grandchildren would live in very similar circumstances.

The only changes that people experienced were bad ones: wars, crop failures, droughts, epidemics and famines. Once the effects of these deadly events wore off, life returned pretty much to the way it had been before... at least for those who were lucky enough to survive.

Humanity lived in a stable state because technological innovation occurred very rarely, and any increased wealth went either to expanding the population or lining the pockets of entrenched elites. Individuals experienced progress, but societies did not.

Geography Constrained Progress In the Past

For most of history, societies were trapped by geographical constraints that made progress impossible. These geographical constraints limited how much food they could produce, the types of food they could produce, and how they could produce them.

Societies evolved technologies, skills and social organizations to acquire the maximum amount of food from their local environment with the minimum level of effort. This quest was so time-consuming that little time was left for innovation in other fields.

Variations in the natural environment led to societies specializing in different methods for acquiring and distributing food. These geographical variations created enormous differences between societies across the globe.

In particular, geographical variations caused variations in the rate at which societies could innovate new technologies, skills and social organizations, and copy the innovations of others. Most of these societies simply could not innovate fast enough or copy enough innovations to produce any real benefits in the people's lives.

No matter what they tried, their efforts did not create progress.

But Then It All Changed

Then within a span of a few lifetimes, billions of people started to experience progress. An elderly European who died in 1920 saw more progress in their lifetime than all the other previous generations of Europeans combined. An elderly person today has seen more progress for the entire world than all of human history combined. It has been a stunning transformation.

Today we live in a period of great progress. This progress has led to a dramatic increase in the standard of living of societies throughout the world. This progress has broadened to such a degree that it is self-sustaining and unlikely to be destroyed by even the most powerful shocks. But progress has affected some people far more than others, leading to great inequalities.

This progress evolved from the bottom up; it was not the creation of any government, institution, leader or set of policies. It is the result of hundreds of generations of people innovating, learning, cooperating and copying to solve local problems. Technology, skills and social organizations all play important roles in this process.

If a person living in late 19th Century Europe or United States had claimed that progress exists, it would hardly have attracted notice. Europeans and Americans at the time widely believed in progress and understood that it directly impacted their lives. Most educated people at the time treated progress as though it were a fact.

Now We Forget How It Used to Be

But then after the impacts of World War I, World War II, the Holocaust, economic recessions and our knowledge of climate change, many people lost faith. A relentless cynicism and pessimism about contemporary life and the future seems to have spread across the Western world. And, most bizarrely, these attitudes are most common among the wealthiest and most educated members of the richest societies that have ever existed.

So we live in a world of unprecedented abundance and progress, but the very people who have enjoyed the most benefits seem to be unaware of its existence. Some even attack the very idea of progress.

How did we get here?

But first, what is progress?

Progress is About the Material Standard of Living

I believe that the most useful definition of progress is "the sustained improvement in the material standard of living of a large group of people over a long period of time." In particular, I focus on changes to the standard of living that are rapid enough and sustained enough that one person could notice positive changes within their lifetime.

Progress within one year that is immediately erased by a regression in the next few years does not qualify as progress. Since a generation is generally considered to be 20 years, I look for relatively uninterrupted progress for at least that length of time. One sharp downturn is not enough to invalidate a decade of progress, but a downturn that lasts for decades surely means that progress did not exist during that time.

Progress is not about enriching a small portion of society. While it is possible to apply the term to changes that exclusively benefit the rich and powerful, I am far more concerned about material progress for the vast majority of citizens.

The Progress-Deniers Are Wrong

Today, a dominant portion of intellectuals in the West doubt that progress exists. They often put the word "progress" in quotes to detach themselves from the idea. Many claim that people are no better off today than they were decades ago. Or they claim that material progress has negative impacts on our soul and psyche that undo the material benefits of progress. Others point to inequality and environmental destruction as proof that progress does not exist.

In this book, I will show that the material standard of living of the bulk of mankind has improved dramatically, but first let me deal with the other claims.

Progress is entirely compatible with environmental destruction. It is also compatible with a healthier environment. Human progress and the natural environment are completely different concepts that are only tangentially related. In this book, I will not deal with the environment.

While there are many who claim that continuing environmental destruction will inevitably lead to the end of progress, the evidence in this book shows that it has not done so yet. And the numerous predictions of the end of progress made in the past have all proven false. Perhaps progress will end sometime in the future, but this in no way changes whether progress exists today or has existed in the recent past.

If you believe that protecting the environment is so important that we must give up on improving people's material standard of living, then you will probably not get much from this book. If you believe that both are important, then read on.

Nor is inequality a valid argument against progress. It is entirely possible for a society to experience progress along with increasing inequality. It is also possible for a society to experience progress with greater equality. Just like environmental destruction, inequality is a completely separate topic from progress. As long as the vast majority benefit from progress, the fact that some people benefit more than others does not invalidate the fact that progress has occurred.

If you believe in equality of outcome so strongly that we should focus on tearing down the successful instead of building up the less fortunate, then you will not get much from this book. I will aim to show that progress has and will in the future help to uplift all classes in society within nations across the world.

Progress is also entirely compatible with bad events. The critics of progress are correct to say that wars, epidemics, famines, depressions, political disorder and other negative events have happened in recent years. They are incorrect in the assumption that the mere existence of these bad events undermines the possibility of progress. I will argue that there has been clear evidence of progress even while these other negative events have taken place.

Progress does not, and indeed cannot, eliminate all problems. Progress often enables us to adapt to problems or lessen their severity. Progress in one area often uncovers problems in other areas. Sometimes those problems are actually more severe than the original problem, and sometimes they are less severe. The more we solve problems, the more noticeable and inconvenient other problems become. This is all true, but, again, this does not mean that progress does not exist.

Progress is not the same as utopia. Indeed, I will argue that the quest for utopia undermines progress. Utopians compare society to an ideal that exists only in their brain. I compare societies with how they were previously and to other societies at the same time period. No matter how much of humanity experiences progress, problems will always exist. Utopia will never be achieved.

Progress is also compatible with entire nations or sub-national groups not being part of it. Certainly, there are communities, cities, regions and countries that have not taken part in progress. It will always be possible to drill down into the data to find examples of groups where progress has not taken place.

But identifying exceptions to the trend does not disprove the trend. More importantly, there have been many examples of populations who experienced no progress in the past suddenly being transformed within one generation. There is no reason to believe that any specific population will be locked out of progress forever.

Let me be clear that I am not endorsing consumerism. Purchasing a flashy new car, clothes or jewelry is not progress. Being able to make visual displays of social status is not a sign of progress. Flaunting social status has been part of human behavior for thousands of years. It will not make you happy or successful.

I am not arguing for consumerism, but I am arguing for materialism. It really does matter whether a person has a flush toilet, clean water, vaccinations, higher levels of education, literacy, access to books or the internet, means of transportation, and a house that protects them from the elements and gives them some personal space. It does matter whether someone can afford to pay for medical treatment that leads to a longer and healthier life. It does matter whether a person can purchase fresh fruit, vegetables and protein year-round. It does matter whether they have a reliable electric grid that powers lights, appliances, computers and cell phones. These things really matter to the quality of a person's life.

I do not make the argument that all ramifications of progress lead to happiness. Rapid change can be disorienting, particularly for older people and those of certain personality types. Humans evolved to survive and reproduce in Hunter-Gatherer bands on the African savanna. Modern society, with all of its progress is not natural to mankind, so it requires adjustment. And many people find that adjustment uncomfortable.

However, I do argue that, as a whole, progress has led to increased levels of happiness. While some parts of progress may promote unhappiness, progress promotes happiness on the whole. Wealthier nations are happier than poorer nations. More affluent people within societies are happier than less affluent people in those same societies. And people living in societies that experience rapid material progress are happier than people living in similar societies that do not experience progress.

This book rarely deals with advances in science. While the gradual expansion of our understanding of our physical and biological world is an amazing achievement (and a clear sign of progress in itself), I will focus on material improvement in the standard of living for the masses. Science has played a role in that progress, but I do not believe that it is the driving force. Scientific advances are more typically the result of progress.

Nor do I deal with technological innovation for its own sake. A specific invention is only a part of progress if it actually increases the material standard of living of the masses. The same goes for institutional innovation. Both are part of progress, but not every innovation automatically leads to progress.

I am not a futurist who speculates about the great positive (or negative) impacts of conceivable innovations. I believe that humans are very good at predicting everything... except the future. Whenever I hear someone making a specific prediction about the distant future, I think that they are seriously overestimating the powers of the human brain to understand a complex reality.

I focus on the present and the past. That is complex enough for me.

I will not dwell on recent events. Progress is not about what happened today, this week, this month or even this year. Within these short timespans, random variations due to sudden events often overwhelm the long-term trend of progress. For this reason, I focus on decades, centuries and even millennia.

This book is not restricted to evaluating the trend in material standard of living of Americans and Western Europeans over the last 50 years. While the wealthiest nations are one part of the overall trend of progress, this book is about the entire world over centuries and even millennia. Which living generation of Westerners (Baby Boomers, Generation X, Millennials, etc.) has lived a better life is far less important than the fact that all of them are far better off materially than 99.999% of all humans who have ever lived.

If you are reading this book and are skeptical of my argument: good for you. Skepticism is essential for testing the validity of ideas.

It is a foundation of knowledge, science and progress. You should not accept my claim just because I assert it boldly. Demand evidence.

But I do recommend asking yourself why you are skeptical of progress. Particularly if you are over the age of 40, examine the changes that have taken place during your lifetime. Imagine if the innovations that took place during your lifetime suddenly disappeared. I believe that, if you can get past nostalgia for how life was during your childhood, you will see enormous changes that have improved your life.

Be sure that your skepticism is not cynicism in disguise. Skepticism is healthy. Cynicism is toxic because it has already decided that things are bad regardless of the facts. Cynicism may not undermine progress, but it does undermine one's ability to enjoy the benefits of the progress that surrounds us.

The Five Keys To Progress

In this book, I introduce the critical concept of the Five Keys to Progress. **If there is any one big take-away from this book, it should be these Five Keys to Progress**.

I believe that the Five Keys to Progress is an essential unifying concept for understanding progress. The Five Keys to Progress are critical because they are the **necessary preconditions** for a society changing from a state of poverty to a state of progress, and they are **actionable** in today's world. In other words, the concept not only helps one to understand the world but also how to make it better.

The Five Keys to Progress enable us to cut through all of the clutter of history and modern times so that we can focus on what matters most. The Five Keys to Progress enable us to answer some of history's most difficult questions, as well as to provide policy solutions and practices that can make the world a better place.

Using the concept of the Five Keys to Progress, it is easier to understand:

- The historical origins of progress.
- Why progress took so long to get started.

- How and why progress started in Northwest Europe.
- How and why progress spread to different societies over time.
- Why so many poor nations were left without progress for centuries.
- Which forces threaten progress today.
- What policies and practices wealthy nations should adopt to keep their progress going.
- What policies and practices developing nations should adopt to enjoy greater progress.

So what are the Five Keys to Progress? To transition from poverty to progress, a society needs to acquire:

1. **A highly efficient food production and distribution system**. This enables societies to overcome geographical constraints on food production so that large numbers of people can focus on solving problems other than getting enough food to eat.

2. **Trade-based cities** packed with a large number of free citizens possessing a wide variety of skills. These people innovate new technologies, skills and social organizations and copy the innovations made by others.

3. **Decentralized political, economic, religious and ideological power**. It is of particular importance that elites are forced into transparent, non-violent competition that undermines their ability to forcibly extract wealth from the masses. This also allows citizens to freely choose among institutions based upon what they have to offer to each individual and society in general.

4. **At least one high-value-added industry** that exports to the rest of the world. This injects wealth into the city or region, accelerates economic growth and creates markets for smaller local industries and services.

5. **Widespread use of fossil fuels**. The incredible energy density of fossil fuels injects vast amounts of useful energy into society, enabling it to solve a wide variety of problems. Without this energy, life would return to the daily struggle for survival that dominated most of human history.

Each of the Five Keys to Progress is necessary for a society to experience progress, but none are sufficient by themselves.

I believe that the degree to which peoples have enjoyed progress is largely determined by long-term historical factors that go back centuries or even millennia. These factors determined the extent to which societies acquired the Five Keys to Progress. For most of human history, there was no progress, because these five key factors were either completely missing or very underdeveloped.

Progress Has Spread Throughout the Globe

Once a society achieves the Five Keys to Progress, it can escape the poverty trap imposed by geography, demographics and politics. Human history can be viewed as a vast evolutionary process that led to the accidental discovery of the Five Keys to Progress. Once these five keys were discovered, they diffused slowly and unevenly throughout the world.

There were six historical breakthroughs that enabled progress to accelerate and diffuse to new parts of the globe:

1. **The emergence of Commercial societies** in Northern Italy about 800 years ago, which combined four of the Five Keys to Progress (productive agriculture, trade-based cities, decentralized power and export industries).

2. **The diffusion of Commercial societies** from Northern Italy to Flanders (modern-day Belgium) and then to the Netherlands and finally to Southeast England.

3. **The migration of Europeans** to much of the rest of the world. The migration of peoples from England to North America was particularly important.

4. **The Industrial Revolution in Britain** in the 19th Century which added the fifth key to progress (widespread use of fossil fuels).

5. **The Allied victory in World War II**, which ended the totalitarian threats of Nazi Germany, Imperial Japan and Fascist Italy.

6. **The collapse of the Soviet Union** in the early 1990s.

Expanding on this list a bit, a specific type of society evolved in the city/states of Northwest Europe between 1200 and 1800. These Commercial societies laid the foundation for our current progress by perfecting four of the Five Keys to Progress.

Because Commercial societies were based upon free peoples with a wide variety of skills living in densely populated cities, they were unusually innovative and willing to copy the innovations of other societies. These societies built more inclusive political, economic and religious institutions that were forced to compete nonviolently against each other to produce benefits for the people.

The Industrial Revolution in the 19th Century radically broadened and deepened this progress by leveraging the awesome energy density of fossil fuels. Fossil fuels enabled the innovation of extraordinarily productive food, energy, transportation and communication technologies. These technologies radically expanded the overall number and variety of other technologies, skills and social organizations. The fifth Key to Progress had finally evolved.

Industrial technologies, skills and social organizations enable people to overcome many, but not all, of the geographical constraints that previously undermined the possibility of progress. For the first time, people anywhere on the globe had the opportunity to experience progress. They could now escape the trap of geographical constraints.

Creating the first society with progress was humanity's greatest achievement. It had taken at least 100,000 years of evolution to do so. Once one society made the jump from poverty to progress, other societies could copy the first one. The genie of progress had been let out of the bottle, and there was almost no way to put it back inside.

When contemporary peoples copy the technologies, skills and social organizations of more successful societies on a grand scale, they can dramatically improve the standard of living and happiness of their people. Japan, South Korea, Singapore, China and India are just a few of the previously poor nations who made a sudden jump to prosperity by copying innovations made in richer nations.

By copying nations who have already achieved the Five Keys to Progress, entire nations can jump from poverty to progress.

Progress Matters to Your Life

Progress is the most uplifting story in human history. It has transformed poverty into prosperity, disease into health, ignorance into education, isolation into connectedness, war and violence into peace and security, slums into housing, and servitude into freedom.

Quite frankly, progress is the single most important force to impact the material existence of humanity. We must protect it in wealthy nations and expand it in developing countries.

Rejecting progress puts serious limitations on people's lives. If you believe that things are bad and that things are getting worse, what are the chances you can have a successful and happy life?

By accepting the reality of progress and learning from the progress of the past, you can give yourself and your loved ones a better shot at living happy and successful lives.

My hope is that this book helps to ramp down the current level of cynicism and to replace it with a new field of inquiry for understanding which factors promote progress. Just as importantly, I hope to spark interest in identifying which actions people need to take to enjoy the benefits of the progress that surrounds them.

So, let's get on to the evidence that progress exists.

MEASURING PROGRESS

In this chapter I will provide evidence that we live in an era of great progress, and that progress benefits the vast majority of mankind. To prove that progress is taking and has taken place, we need objective means by which to measure it. Just pointing to a few examples where progress takes place is no more definitive than pointing to a few examples where progress does not take place. We must look for overall trends.

Unfortunately, there is no one metric that accounts for all dimensions of progress. So instead, I will take a broad approach by using many different development metrics. These metrics include measures of economic growth, poverty, agricultural production, diet, sanitation, drinking water, life expectancy, neonatal mortality, education, housing, happiness and more. I deliberately cast a broad net in order to capture as many dimensions of material well-being as possible.

One of the key problems with documenting progress is finding good metrics that both go far enough back in time and that cover the entire world. Not surprisingly, there is far more data related to recently industrialized nations than those same societies in, say, 1500. Nor is it surprising that data is far easier to acquire for wealthy nations than poor nations. In many cases, I need to use different methods for different time periods and different nations. While not exactly comparing apples to oranges, it is a bit like comparing Gala and Red Delicious apples to Fuji apples. This is less than ideal, but it is hard to do otherwise.

The metrics that I use come from a wide range of official government and NGO sources. I present the metrics in a series of graphs. Unfortunately, the scope of the data is not easily digestible in static graphs. If you would like to inspect the data in more detail, I encourage you to explore a website that has spent a great deal of time gathering important metrics: Our World in Data.

Since there are over 200 nations today, it is not realistic to examine development metrics for every one of them in this book. And averages can cover up variations between rich and poor nations. We need a way to narrow the sample to a manageable number, but not in a way that creates a distorted impression of overall trends. In order to ensure that the data covers a very broad segment of the world's population, I decided to focus on four distinct categories of nations.

The first group, which I will call the "Wealthy 12", consists of 12 Western nations that industrialized early and currently have very high standards of living. Those nations are the United States, United Kingdom, Australia, Belgium, Canada, France, Germany, Netherlands, New Zealand, Norway, Sweden, and Switzerland. The Wealthy 12 gives us a good overview of the trends within the wealthiest nations.

The second group that I will show data for is what I call the "Populous 12". This group consists of 12 of the most populous nations that did not have high per capita GDP in 2020. This group consists of China, India, Brazil, Congo, Egypt, Ethiopia, Indonesia, Iran, Mexico, Nigeria, Pakistan, and Turkey. Together these nations make up 58% of the world's population and cover every continent except Australia and Antarctica. The Populous 12 gives us a broad overview of trends for people who live outside the wealthiest nations.

The third group is what I call the "Bottom 20". This group consists of the 20 nations with the lowest scores on the United Nations Human Development Index in 1990 (the earliest year available). The nations in this group consist of Afghanistan, Benin, Burma, Burundi, Central

African Republic, Congo, Gambia, Guinea, Malawi, Mali, Mauritania, Mozambique, Niger, Papua New Guinea, Rwanda, Senegal, Sierra Leone, Sudan, Tanzania and Uganda. The Bottom 20 gives us a good overview in trends of the most desperately poor nations in the world. If there is any group of nations that should lack evidence of progress, it is these 20 nations.

The last group of nations is what I call the "Transformative 16". This group consists of nations that experienced at least one generation of very strong economic growth after 1950 (or 20+ years of per capita GDP growth of over 3 percent). This level of economic growth would lead to a doubling of the standard of living of their people within one generation.

The Transformative 16 includes representatives from many different regions and cultures: Spain, Ireland, Japan, Hong Kong, Taiwan, Thailand, Singapore, South Korea, Indonesia, China, India, Israel, Botswana, Trinidad, Puerto Rico and Chile. The Transformative 16 gives us a good overview of the nations that experienced the fastest economic growth. It tests whether very rapid economic growth translates into positive changes throughout society.

Together, the Wealthy 12, the Populous 12, the Bottom 20 and the Transformative 16 are a solid set of groups with which to test whether the world has experienced widespread progress over the last few decades. Given the breadth and diversity of the four groups it seems unlikely that any broad trends will be missed by narrowing the sample down from all nations in the world. In some cases, I will supplement these four groups with other data that seems relevant.

WARNING: Proving that progress exists requires lots of data in graphs. Some people love to dig into data, while others hate it. If you love data, this chapter is for you. If you hate it, you can skim over a few of the sections to understand the main points and then read the conclusion at the end of this chapter.

A few book-keeping notes:

Each metric varies greatly in the number of nations and years due to the availability of data. I have done my best to be as comprehensive as possible given data that is accessible on the internet. In order to facilitate visual inspection, I have added an "Average" displayed in a thick black line wherever there are more than four nations with data. This average is sometimes jagged when individual data points are missing.

Wherever data permits, I will focus on women rather than the total population to show that progress has not excluded them. Note also that the use of these categories leads to some double counting. China, India and Indonesia are in both the Populous 12 and the Transformative 16. Congo is in both the Populous 12 and the Bottom 20.

Additional full-color line graphs for all the countries and their metrics can be found at my website. See the front matter of this book for a URL and a QR code to the website.

Per Capita GDP

Economic growth is central to progress. With economic growth, it is possible to pay for education, health care, transportation, housing and other factors that promote an increased standard of living. Without economic growth, progress becomes far more difficult because there are simply not enough resources.

In this study, I will measure economic growth using per capita GDP in real 2011 dollars because this takes into account population and inflation. Data on per capita GDP since 1950 are widely available.

In addition, Angus Maddison has created a publicly accessible database of estimates of per capita GDP across the world going back to 1 AD. Maddison started with the current per capita GDP, and assigned a figure of $450 income per year to the poorest of societies in history. This is roughly the level of economic activity needed to support basic human survival and reproduction. He then worked backward by using available economic data to make estimates for every society in the world going back to 1 AD. These are, of course, estimates, but they do give us a rough order-of-magnitude level of the standard of living of people throughout the last 2,000 years.

For the Wealthy 12 nations, one can see the power of exponential growth of per capita GDP. From year 1 to the year 1000, the "curve" is a virtual flat line. If we had data from before year 1, we would see that this flat line had existed for millennia.

But within that apparent stagnation was very gradual innovation feeding upon itself. Somewhere around the year 1500, the rate of innovation had accelerated enough to produce an improvement in people's lives. Since 1820, the curve has continually become steeper making the exponential nature of growth obvious.

The curve gets particularly steep after 1950 (when the data is no longer based upon estimates). With the exception of Switzerland, all nations in the Wealthy 12 had a per capita GDP of $16,000 or less (in current values) in 1950. This level of income is less than the current level for Mexico. By 2017, every nation in the Wealthy 12 had a per

Per Capita GDP

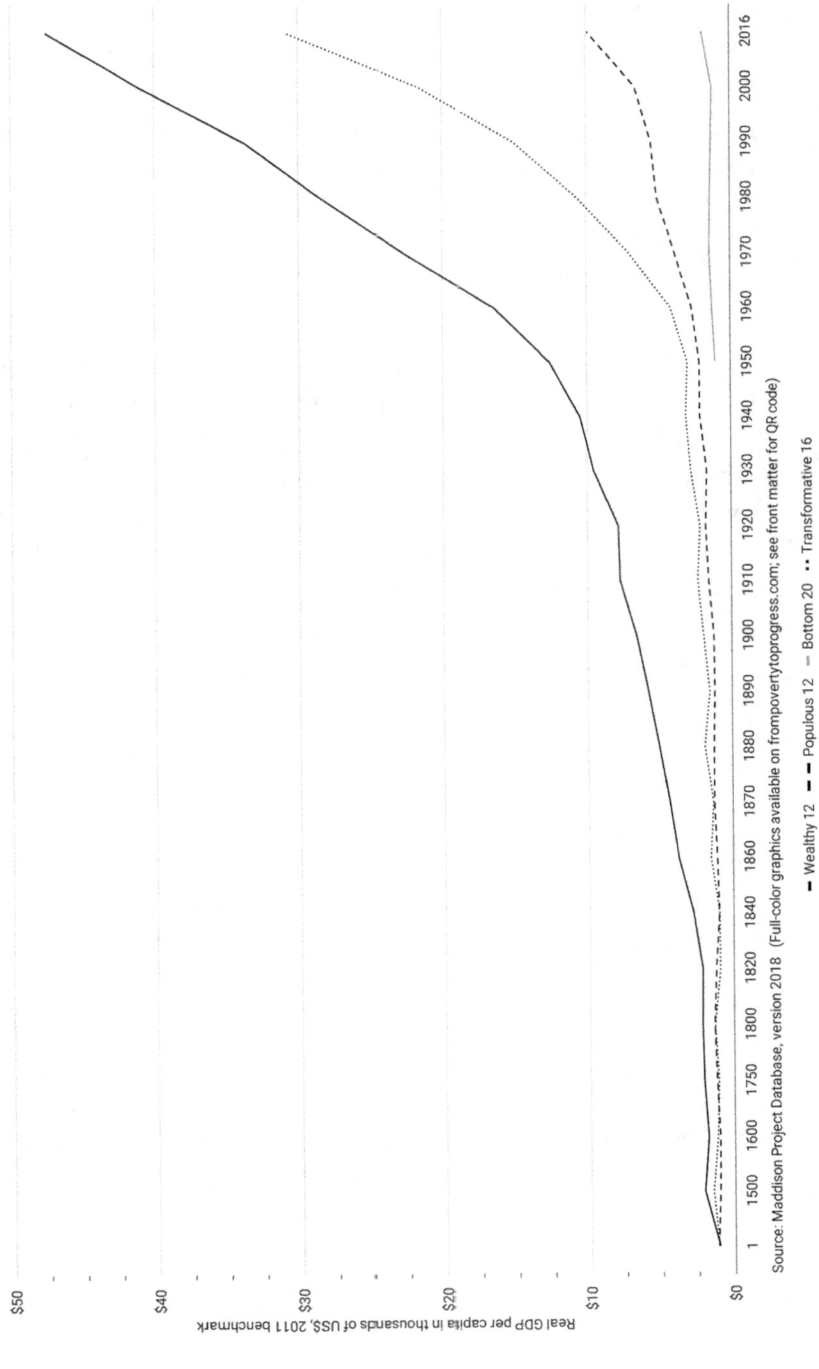

Real GDP per capita in thousands of US$, 2011 benchmark

$50
$40
$30
$20
$10
$0

1 1500 1600 1750 1800 1820 1840 1860 1870 1880 1890 1900 1910 1920 1930 1940 1950 1960 1970 1980 1990 2000 2016

Source: Maddison Project Database, version 2018 (Full-color graphics available on frompovertytoprogress.com; see front matter for QR code)

— Wealthy 12   - - Populous 12   — Bottom 20   •• Transformative 16 capita GDP of $37,000 or more, more than double the previous levels. On average, these nations quadrupled their per capita income between 1950 and 2016.

For the Populous 12, we can see a similar exponential curve, but the levels of income are much lower. The upward trajectory of the curve did not really start until 1950. The increase in the steepness of the curve has been particularly strong since the year 2000.

In 1950 all the nations in the Populous 12 had levels of per capita GDP much lower than the Wealthy 12, with Mexico, being the highest at $4,179, and China being the lowest at $637. By 2016, all but Ethiopia and Congo had surmounted the level of Mexico in 1950. Some experienced long, slow economic growth with some important dips — Mexico, Brazil, Turkey, Iran and Egypt. Other nations saw spectacular growth after 1980, with China being the premier example.

Even some of the laggard nations within the Populous 12 experienced a transformational change in levels of per capita GDP. Pakistan quadrupled from $1,258 to $5,223, and Nigeria more than doubled from $1,961 to $5,360. Even Ethiopia more than tripled from $520 to $1,635.

Unfortunately, the fact that four of the Populous 12 are oil-exporting nations gives a distorted view of their actual economic growth. Many of these nations made far more money from exporting oil than from other products. And the profits from oil exportation tend to go to politically connected elites, so it is unclear how much this economic growth in oil-exporting nations actually benefitted the masses.

The only nation among the Populous 12 that fits the idea of the "poor getting poorer" is civil-war-ravaged Congo, which declined from $1,641 to $808. Obviously, the Congo has not experienced anything like progress over the last few generations. If most nations had such low and declining levels of per capita GDP, this would clearly invalidate the progress hypothesis. Fortunately, very few nations have experienced such declines.

The pattern for the Bottom 20 nations is quite different to the previous two groups. Economic stagnation persisted in those countries until around 2000. Unfortunately, we have no data from before 1950, but there is every reason to believe that levels of per capita GDP among the Bottom 20 nations were very low.

There were some very slow increases in per capita GDP before 1970, but then that progress was erased in the following three decades. It was during this period that it became mainstream thinking to see the poorest nations in the world as being trapped in poverty and beyond redemption. At the time, many believed that only wealthy Western nations could experience long-term economic growth.

After 2000, however, even the Bottom 20 began to experience real economic growth, perhaps for the first time in their history. The strongest economic growth was in Burma, which more than sextupled its per capita GDP, and Sudan, which doubled it. The other nations experienced much slower economic growth, while Niger and Malawi stagnated. Only civil-war-torn Congo saw negative growth during this period.

Of course, an upward trend for the past two decades is not a very long-term trend. Whereas it appears that the Wealthy 12 and Populous 12 have experienced long-term self-sustaining increases in their standard of living, it is too early to declare victory for the Bottom 20. Guarded optimism, however, is in order. Based on all the trends that we have seen for other nations, it seems likely that the Bottom 20 are already reaching or ascending the steep section of the exponential curve.

Exponential growth is even more obvious among the Transformative 16. Those nations all experienced rapid economic growth since 1950 after having suffered millennia of stagnation. This should not be surprising given that this group was selected because of their high levels of per capita economic growth over a long period of time.

The changes among the Transformative 16 are stunning nonetheless. In 1950 the wealthiest nation was Ireland, which had a per capita GDP of $6,983. This made Ireland one of the poorest nations in Western Europe.

By 2017 all the nations in the Transformative 16 had reached levels higher than Ireland had in 1950, except India, which was just below that level. Most of the Transformative 16 had per capita GDP six times what Ireland had in 1950.

Given that some of these nations are some of the most populous in the world, this is a stunning transformation. Virtually all of the nations in the Transformative 16 had levels of per capita GDP that greatly exceeded the levels of the Wealthy 12 in 1950: only India was lower with Indonesia and China at their level. And given the current trajectory, there is every reason to believe that rapid economic growth will continue in the future.

Of all of the metrics used in this study, per capita GDP is the one that leads to the most varied outcomes. Despite this, there was clear progress throughout the world. About half the nations experienced strong economic growth that transformed their people's standard of living. A handful of oil-exporting nations experienced economic growth that may not have affected their people positively. And the Bottom 20 saw nothing but stagnation until 1990 and then started to experience economic growth afterward.

Absolute Poverty

Of course, the poor may have missed out on this economic growth. Perhaps the rich have monopolized all the economic gains. So we must go beyond per capita GDP to see if the bottom of the economic scale enjoyed the benefits of economic growth.

First, let's look at the trends for absolute poverty from 1800 to today. Absolute poverty measures the percentage of people who live on $1.90/day or less (this amount is adjusted for inflation and cost-of-living). $1.90/day is a level of poverty that is almost impossible to imagine for a modern-day citizen of a Western nation. Sadly, it is roughly the standard of living that the vast majority of humans have had for most of history.

In 1800 89.7% of the world's population lived in a state of absolute poverty. The lucky people who lived above this pathetic level of existence were overwhelmingly concentrated in Northwest Europe and the United States. Even as late as 1970 a majority of the people on planet Earth lived on less than $693/year.

The drop in absolute poverty over the last 30 years has been nothing short of revolutionary. Poverty rates have dropped from 40.2% to 10.1% — a quarter of their previous level. Within a little over 200 years, the world transformed from virtually all people living in absolute poverty to very few of them doing so.

In East Asia, the progress in reducing absolute poverty has been nothing short of astounding. Today we forget just how poor East Asia was only a few generations ago. Most East Asians in 1970 had much the same standard of living as they had had 2,000 years previously.

In 1970 East Asia had levels of absolute poverty rates that were roughly the same as Sub-Saharan Africa at the time. By 2006, the absolute poverty rate in East Asia had dropped to 1.7%! In less than two generations East Asia had virtually eliminated absolute poverty.

If we look at the overall distribution of world income, the changes are just as clear. In 1830 virtually the entire world lived below the absolute poverty line. By 1939 Europe and North America had grown far richer, but the rest of the world was still trapped in poverty.

By 1970 we lived in a bifurcated world: a wealthy First World consisting of Europe, North America and Japan, and a desperately poor Third World. The First World was rich, the Third World was poor, and few people were in between. Today many people believe that we still live in that world.

By 2020, though, rapid economic growth had transformed the Third World so that many parts of it began to resemble the First World. Today, absolute poverty is rare. Most of the people who Westerners perceive as poor in other nations are middle income. And more poor people move up to middle-income status virtually every year.

Even if there had been no other positive trend over the last 50 years, it would be hard not to consider this proof enough of progress during our time. Of course, global trends can cover up national counter-trends, so we must dig deeper into the data.

Absolute Poverty (living on less than $1.90/day)

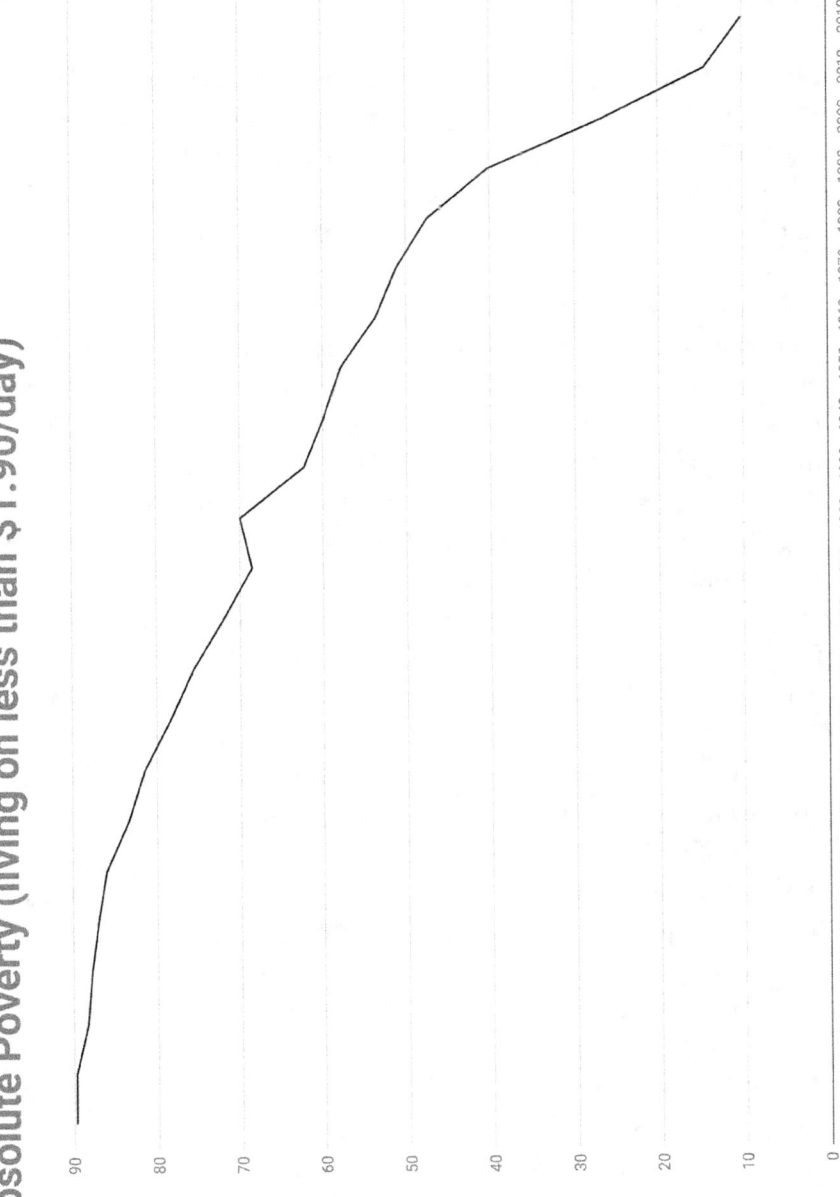

Percent of world population living on less tf an $1.90/day (adjusted for inflation and price differences)

90 80 70 60 50 40 30 20 10

1800 1810 1820 1830 1840 1850 1860 1870 1880 1890 1900 1910 1920 1930 1940 1950 1960 1970 1980 1990 2000 2010 2019

Poverty Rate by Nation

Another means for measuring progress in reducing poverty is the Poverty Headcount Ratio at $3.20/day. This metric measures the percentage of the total population in each nation that earns under $3.20/day or $1168/year. This amount is indexed for inflation and cost-of-living to make the comparison between nations and over time legitimate, so all values are given in today's values. For the Wealthy 12, the number of poor by this measure is so low as to make the measurement meaningless.

Among the Populous 12, the average poverty rate declined dramatically from 62% in 1984 to 14% in 2014 — a quarter of its previous level. China reduced poverty from 90% in 1990 to 7%. Indonesia reduced poverty from 92% to 27%. Mexico's figure dropped from 23% to 11%. Bangladesh, Pakistan and Turkey all made substantial progress by this measurement. Of the populous low-income nations, only Nigeria failed to make real progress (although the data is very sparse for this nation).

For the Bottom 20, the data is also encouraging. The average declined from 83% in 1989 to 71% in 2013. Most nations have made substantial progress, with Mauritania (from 66% to 24%) and Gambia (from 87% to 38%) showing the most impressive results. The only country where the level increased was in Malawi, but some others showed only slow progress: Congo, Burundi, Papua New Guinea and Senegal.

Part of the problem is that these nations started out so poor that this metric misses some important progress. When we use $1.90/day instead, the progress is more evident in the Bottom 20. Using this metric, every nation within the Bottom 20 shows real progress, with Rwanda being the exception (dropping from 61% to 56%).

Of course, it is hard to miss the fact that poverty in the Bottom 20 is still at appalling levels and far more progress needs to be made. Given the trends over the past 30 years, however, there is reason to be optimistic about continued progress in the future.

The Transformative 16 experienced very significant declines in

Poverty Rate (at $3.20/day)

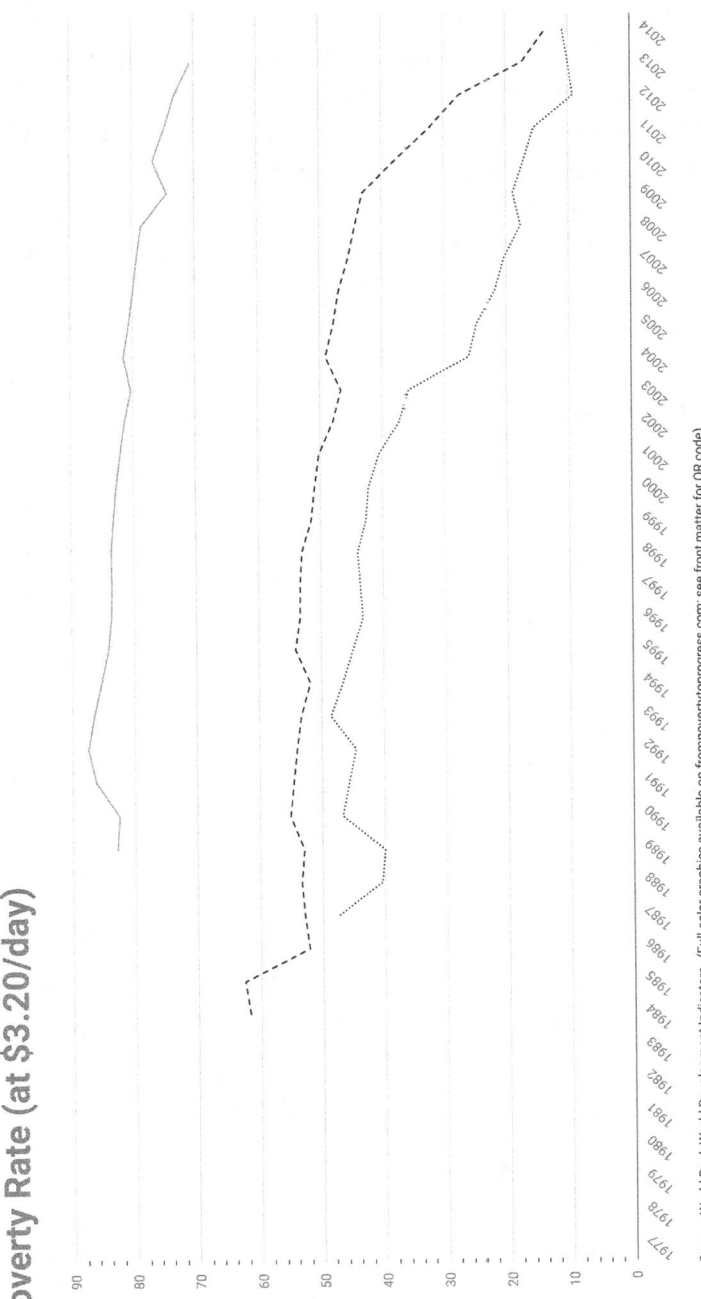

Poverty percentage (at $3.20/day adjusted for inflation and PPP)

Source: World Bank World Development Indicators   (Full-color graphics available on frompovertytoprogress.com; see front matter for QR code)

■ ■ Populous 12   ▬ Bottom 20   •• Transformative 16

poverty. The average dropped from 47% in 1987 to 11% in 2014 — less than a quarter of its previous level. Except for Spain and Ireland who already had very low rates of poverty by this measure, and Trinidad for which we have little data, all the nations had very large drops. Chile and Thailand virtually eliminated poverty, while China, India and Indonesia saw dramatic drops. By 2017, only India still had a significant problem with poverty although the levels were dropping fast.

UN Human Development Index

Robust levels of economic growth and dramatic declines in poverty are not enough to prove that progress exists. We also need to look at broader indexes of material standards of living.

The Human Development Index (HDI), which was created by the United Nations, combines several development metrics into one index. It is now the most widely used index for measuring material standard of living. The components of HDI are life expectancy at birth, expected years at school, mean years of schooling and Gross National Income per capita (adjusted for purchasing power). The index ranges from 1.00 (the highest possible score) to 0.00 (the lowest possible score).

HDI gives us reliable data for measuring progress from 1990 to the present. In 2018, the index values range from first-placed Norway at 0.954 to 189th- place Niger at 0.377. For comparison purposes, the United States and the United Kingdom are tied for 15th at 0.92.

The Human Development Index overwhelmingly confirms the fact that there has been widespread progress throughout the world for the last 30 years. Between 1990 and 2018, the World's HDI improved from 0.598 to 0.731 (a 22.2% improvement).

For the Wealthy 12, every nation improved its HDI over the last 30 years. The United States (+7%), Canada (+9%) and Australia (+8%) showed the least progress, while the UK (+19%) and Germany (+17%) showed the most progress. It is important to note that the score for every one of the Wealthy 12 improved by less than the world average.

For the Populous 12, every nation improved its HDI over the last 30 years. Even the nations that improved the least still made a substantial improvement: Mexico (+18%), Congo (+22%) and Brazil (+24%). Except for Congo, these were the wealthiest nations within this group in 1990. The most impressive improvements were made in China (+51%) and India (+50%). As is common with many of the metrics in this book, the two most populous nations in the world have been setting the pace for progress over the last 30 years.

For the Bottom 20, every nation improved their HDI over the

last 30 years, and all but Congo (+22%) and Central African Republic (+19%) improved more than the global average. And those two countries showed improvements that were only slightly below average. Indeed, only 6 of the Bottom 20 nations in 1990 failed to double the world average for improvement (+22%). Many of the nations that were frightfully underdeveloped in 1990 showed spectacular progress: Niger (+78%), Mozambique (+106%), Mali (+85%), and Rwanda (+119%).

The Transformative 16 all showed substantial improvements in HDI from 1990 to the present. India had the lowest level (0.427) in 1990. By 2017, India had increased by 50% to 0.64. Seven of the Transformative 16 had HDI in 2017 that met or exceeded the levels of the Wealthy 12 in 1990.

Because of the comprehensive nature of the HDI data, we can investigate in greater detail than for other metrics. Every single geographical region showed significant improvement. The slowest improvements were in Europe and Central Asia (19%), while the largest improvements were in South Asia (46%).

Most importantly, the poorest nations showed the most improvement. Very High Human Development countries (mainly in Europe and North America) improved only 15%, and OECD nations improved a similar 14%. Both of these figures, while better than in 1990, are far below the improvement in the rest of the world.

By comparison, Sub-Saharan Africa improved by 35%, Low Human Development nations improved by 44%, and the Least Developed Countries improved by 51%. These are all astounding improvements for regions that saw no evidence of progress for thousands of years!

If we look at how individual nations progressed in comparison to each other, the entire world seems to be progressing. In the graphic above nations are color-coded by which quartile (25%) they belonged to in 2018. The nations with dark green lines are in the top quarter in 2018, while the nations in orange lines were in the bottom quarter in 2018. The light green and yellow lines are the middle quarters.

Viewing this graphic, it is obvious that virtually all nations improved their HDI index between 1990 and 2018. Just as importantly, if one

UN Human Development Index

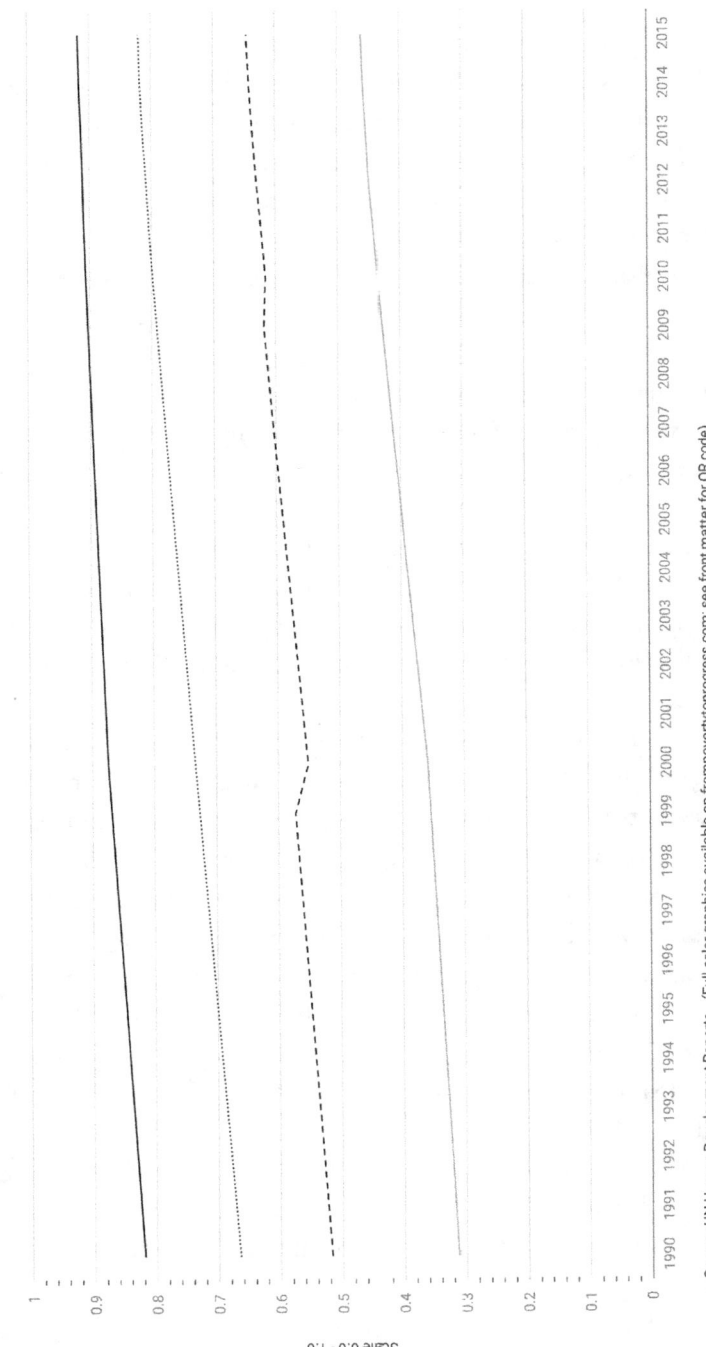

Scale 0.0 - 1.0

Source: UN Human Development Reports   (Full-color graphics available on frompovertytoprogress.com; see front matter for QR code)

Wealthy 12   Populous 12   Bottom 20   Transformative 16

inspects the colors carefully, one can see that the nations remained in roughly the same order. It is as if all nations are going up an escalator with the highest HDI nations getting on first and the lowest HDI nations getting on last. All nations are going up, but some nations got a head start. Most importantly, the destinations are the same: a significant improvement in HDI scores.

There are some exceptions. Syria started in the middle (0.59) in 1990, but then started to drop dramatically in 2012 when their civil war started. Lesotho also dropped substantially until 2006 before improving again. Turkey dramatically improved. Despite these exceptions, there is a remarkable similarity in the improvement for all nations between 1990 and 2018. The poorer nations are closing the gap, but rarely surpassing richer nations in rank.

When the focus shifts to individual nations, the trend is even more apparent. Of the 145 nations with complete data since 1990, all but one improved (Civil War-torn Syria had a decline of 1.6%). And even Syria showed clear improvement from 1990 until 2009, just before the violence started.

If we compare where nations started in 1990 and how much they improved over the next 30 years, it is clear that the poorest nations prospered the most. Among the 20 nations that improved the least in the last 30 years, only four (Syria, Ukraine, Namibia, and tiny Eswatini) were below the world average (0.598) in 1990. None of the 20 nations who improved the most was already above the world average in 1990.

And the pattern remains the same if we restrict our view to the most populous nations. All of the most populous nations with below-average levels of development in 1990 improved more than the world average: China (51%), India (50%), Pakistan (39%), Indonesia (35%), and Bangladesh (58%).

Of course, there are laggard nations, but the number of them is surprisingly small. Only 12 of 145 nations were below average in levels of development in 1990 *and* improved at a rate that was lower than the world average.

While there has been clear progress for almost all nations since

1990, it is important to note that very few nations jumped up or fell back very far in rank by 2018. Among the top 40 in 1990, only two nations (Brunei #42 and Barbados #50) dropped out of that category. Among the bottom 40 in 1990, only four moved up out of the category, with Rwanda, which moved up to #122 being the biggest jump.

Only six nations (Turkey, Iran, China, Singapore, Thailand and Ireland) moved up more than twenty spots between 1990 and 2018. Turkey's move of 33 spots from #86 to #53 was the largest jump.

Alternatively, only eight nations (Libya, Ukraine, Moldova, Syria, Tonga, Kyrgyzstan, Lesotho and Tajikistan) fell more than 20 spots between 1990 and 2018. Not surprisingly, many of these nations have been rocked by civil wars or political turmoil during this period.

Far more typical are the 89 of 145 nations that were within ten spots of their 1990 ranking in 2018. Based on the HDI Index, one of the best measurements of progress available, almost all nations experienced progress but there were relatively small differences in their rankings in 2018 compared to their ranking in 1990.

It should also be noted that the top of the list is just as heavily weighted towards Europe and nations settled by Europeans in 2018 as it was in 1990. While Ireland and Singapore entered the Top 20 by 2019 and the UK and Liechtenstein dropped out, the Top 20 in 1990 was quite similar to the Top 20 in 2018. What had happened is that the rest of the world has narrowed the gap considerably.

Literacy

Literacy is one of the keys to success in modern life. Throughout the bulk of history, literacy either did not exist or was restricted to a very small elite. Most recently, literacy has improved to the extent that it has become nearly universal.

In this study, I will use literacy rates for young women aged 15 to 24 as a metric. This is a better standard than overall literacy because older persons who grew up in an earlier period have lower literacy rates because they were educated before much of the progress took place. And I will use women, instead of men and women combined, because low rates of literacy among women have traditionally been a more serious problem.

Literacy rates for the Wealthy 12 were near 100 percent in 1970, so it was hard for any additional progress to be achieved. For the Populous 12, literacy rates for young women have improved for every nation. In 1981 the average literacy rate was 63%. By 2014, it had grown to 94%.

A number of these nations are now approaching 100% literacy: China, Brazil, Mexico, Turkey, Indonesia and Iran. In addition, Congo, India and Egypt are making such rapid progress that it seems likely that they will achieve similar levels soon.

Only Mexico was above 90% literacy in 1980. Today only Ethiopia and Pakistan are far below 80% literacy, but both of these nations have shown important progress.

For the Bottom 20, literacy for young women has improved. The average increased from 23% in 1979 to 69% in 2014 — tripling the previous level. For nations that have data over a significant period, all show important progress. Sudan, Mauritania and the Central African Republic show downward trends, but data is sparse on the first two nations, so it is hard to know if the numbers reflect a real trend.

If one compares data before 1980 to recent data, the trend is clear. Before 1980, all nations with data except Rwanda had literacy levels for young women under 30%. After 2014, all nations with data have levels over 35%.

Literacy Rate for Young Women Aged 15-24

Percentage of all young women aged 15-24

Source: UNESCO (Full-color graphics available on frompovertytoprogress.com; see front matter for QR code)

▬ ▬ Populous 12   ▬ Bottom 20   ⋯ Transformative 16

Most of the progress in increasing literacy within the Transformative 16 nations occurred before the period for which we have solid data. The Transformative 16 had very high levels of literacy for young women in 1980 (the earliest year on record) and maintained those levels.

The only three nations that had less than 96% literacy in 1980 — China, Indonesia and India — made significant progress in the following years. China and Indonesia moved up to near 100%, while India increased from 40% to 82%. There is every reason to believe that all nations in the Transformative 16 will be near 100% literacy soon, putting them on a par with the Wealthy 12.

Sanitation

Sanitation systems are one of the most important public health innovations in world history. Nobody likes to live surrounded by human waste, and the failure to avoid this is life-threatening. In measuring levels of sanitation, I will use the percentage of all persons with access to improved sanitation.

According to the World Health Organization, "Improved sanitation includes sanitation facilities designed to hygienically separate excreta from human contact." The flush toilet and pit latrines with covered pits are two of the simplest technologies that qualify. This data is generally available from 1990 to 2015.

For the Wealthy 12, almost universal access to improved sanitation was achieved before 1990, so improvement from that level was impossible. For the Populous 12, important increases in the access to improved sanitation took place. In 1990 45% of the citizens of Populous 12 nations had access to improved sanitation facilities. This improved to 67% in 2015.

Some nations had relatively high levels of access in 1990 but still managed to improve: Turkey, Egypt, Iran, Brazil and Mexico. All the other nations, except Nigeria, which declined, and Congo, which showed only modest progress, showed important progress. China, Indonesia, Pakistan and Ethiopia showed the most progress.

Among the Bottom 20, access to improved sanitation improved, but at a slower rate. In 1990 19% of the people had access to improved sanitation. By 2015 the rate had almost doubled to 34%.

Burma, Rwanda and Mauritania showed the most progress, while only Gambia declined (although it had the highest level in 1992). Overall, there are clear signs of progress, although the levels are much less than we see for the Populous 12. In almost every country, the levels are under 50% access to improved sanitation and many are still below 20%.

For the Transformative 16, most had achieved very high levels of access to improved sanitation before the period for which we have data. Among the three nations that have relatively low access to improved sanitation — China, India and Indonesia — all made substantial progress by 2015. The average increased from 79% in 1990 to 88% in 2015. Today, only India has access levels under 60%. While India still has a long way to go to get to 100%, the positive trend is clear and there is no reason to expect the progress to be interrupted.

Access to Improved Sanitation Facilities

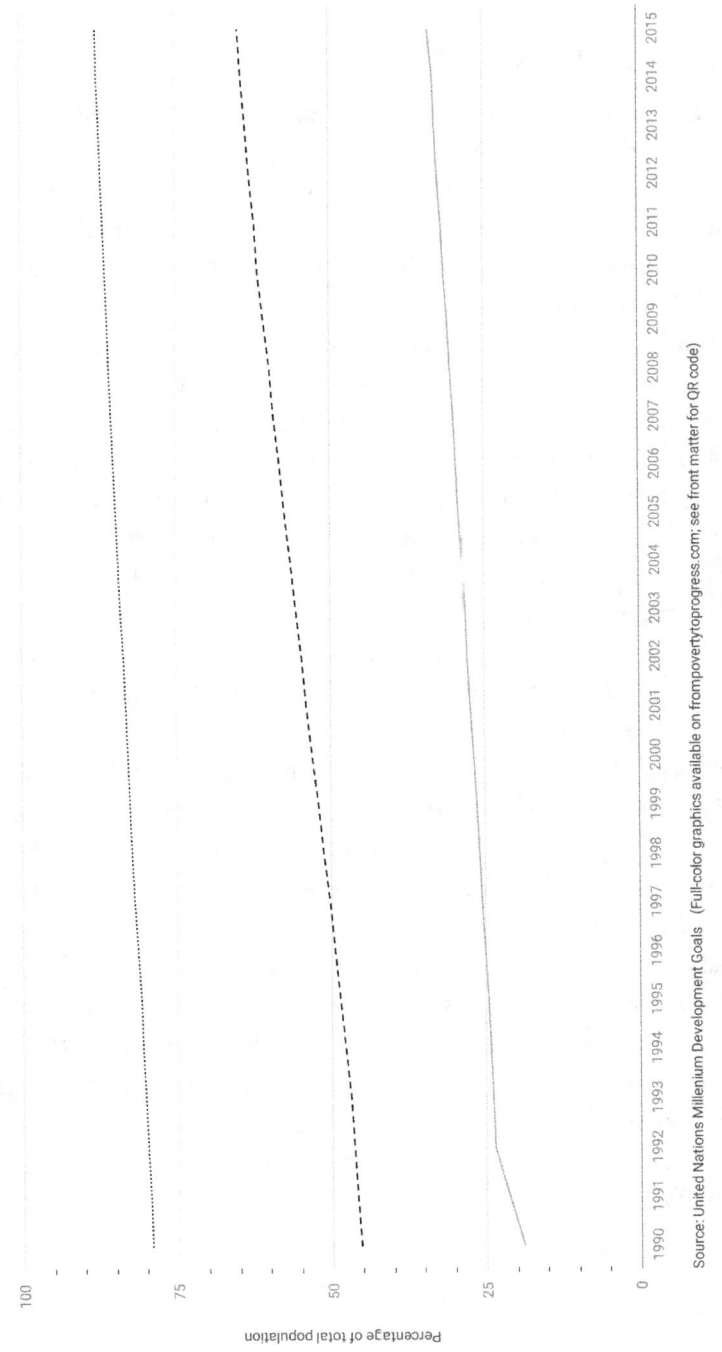

Percentage of total population

Source: United Nations Millenium Development Goals (Full-color graphics available on frompovertytoprogress.com; see front matter for QR code)

■ ■ Populous 12   ─ Bottom 20   ▪ Transformative 16

Drinking Water

The general trend for the percentage of the population with access to improved drinking water sources is largely the same. According to the World Health Organization, "Improved drinking water source is a source that, by nature of its construction, adequately protects the water from outside contamination, in particular from fecal matter. Common examples: piped household water connection."

The Wealthy 12 had already reached or were near 100% by 1990 so there is no change. For the Populous 12, the average percentage of the population with access to improved drinking water increased from 69% in 1990 to 86% in 2015. Every nation but Nigeria, Ethiopia and Congo had reached a level of least 85% in 2015. In addition, every nation but Congo was either above 85% in 1990 and improved slightly or was below this level and improved substantially. There is every reason to expect the trend to continue to improve in future years.

For the Bottom 20, the improvement in access to drinking water was significantly better than for sanitation. In 1990, the average was 49%, while it increased to 69% in 2015. Nine of the Bottom 20 had reached levels of over 85% by 2015, while only one of them (Gambia) had achieved this level in 1990. The only three nations that did not show strong progress were Papua New Guinea, Congo and Tanzania. Except for these three nations, there is every reason to be optimistic about future trends.

The Transformative 16 either had very high levels of access to improved drinking water sources in 1990 or made substantial progress towards that goal afterward. In 1990, the average was 89%, but it reached 97% in 2015. As we have seen in other metrics, China, India and Indonesia were laggards in 1990, but they all made significant gains in the following years. China and India had almost reached the levels of the Wealthy 12 by 2017, while Indonesia is on track to do so soon.

Access to Improved Drinking Water Sources

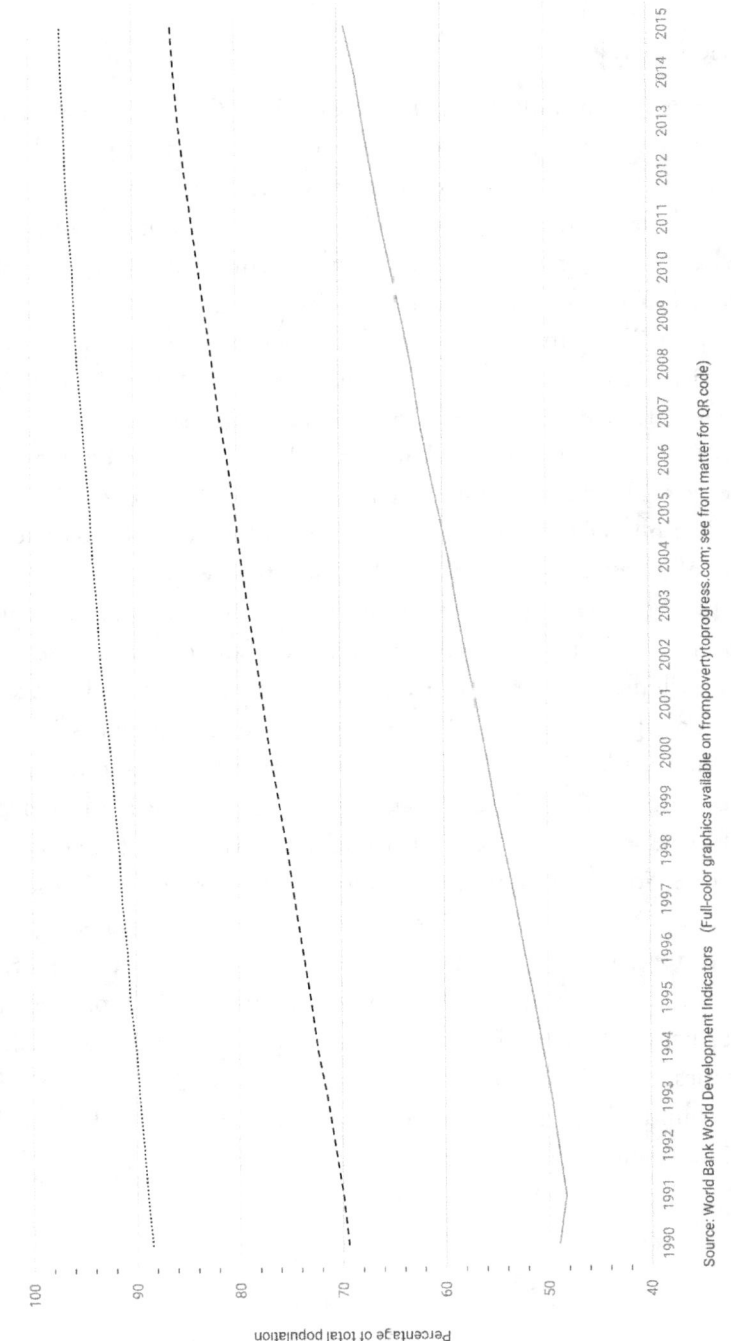

Percentage of total population

1990 1991 1992 1993 1994 1995 1996 1997 1998 1999 2000 2001 2002 2003 2004 2005 2006 2007 2008 2009 2010 2011 2012 2013 2014 2015

Source: World Bank World Development Indicators   (Full-color graphics available on frompovertytoprogress.com; see front matter for QR code)

■ ■ Populous 12   — Bottom 20   ∙∙ Transformative 16

Electricity

Widespread use of fossil fuels is one of the Five Keys to Progress. Perhaps the most important means to access energy derived from fossil fuels is the electrical grid. With access to the electrical grid, people can light their homes, and power electrical appliances, computers and mobile devices. Perhaps no other factor separates a modern society from a traditional society as clearly as a robust electrical grid.

We will use the percentage of the total population with access to electricity from 1990 to 2014 as our metric. Before this period, the Wealthy 12 had already long achieved 100% access.

With the Populous 12, all nations achieved progress by this metric. The average increased from 65% in 1991 to 83% in 2014.

Nations such as Brazil, China, Iran and Mexico already had high rates of access to electricity in 1990, and they were all near 100% by 2014. The remainder of the nations within the Populous 12 achieved substantial progress. With the exception of Congo and Ethiopia, all of them will probably reach 100% access to electricity with a few decades.

Among the Bottom 20 nations, access to electricity has improved at a slower rate. In 1990 all but two nations (Burma and Sudan) had access under 20% and the average was a mere 11%. By 2014 the average had increased to 29%, and only civil-war-torn Sierra Leone failed to improve. So, while the Bottom 20 has a long way to go to give their people access to electricity, important steps have nonetheless taken place.

Among the Transformative 16 nations, every nation increased access to electricity from 1990 to 2014. Most of their progress took place before the period for which we have data, but Botswana, India, Indonesia and Thailand all dramatically increased access after 1990. One hundred percent access to electricity seems achievable for these nations within a few decades.

Access to Electricity

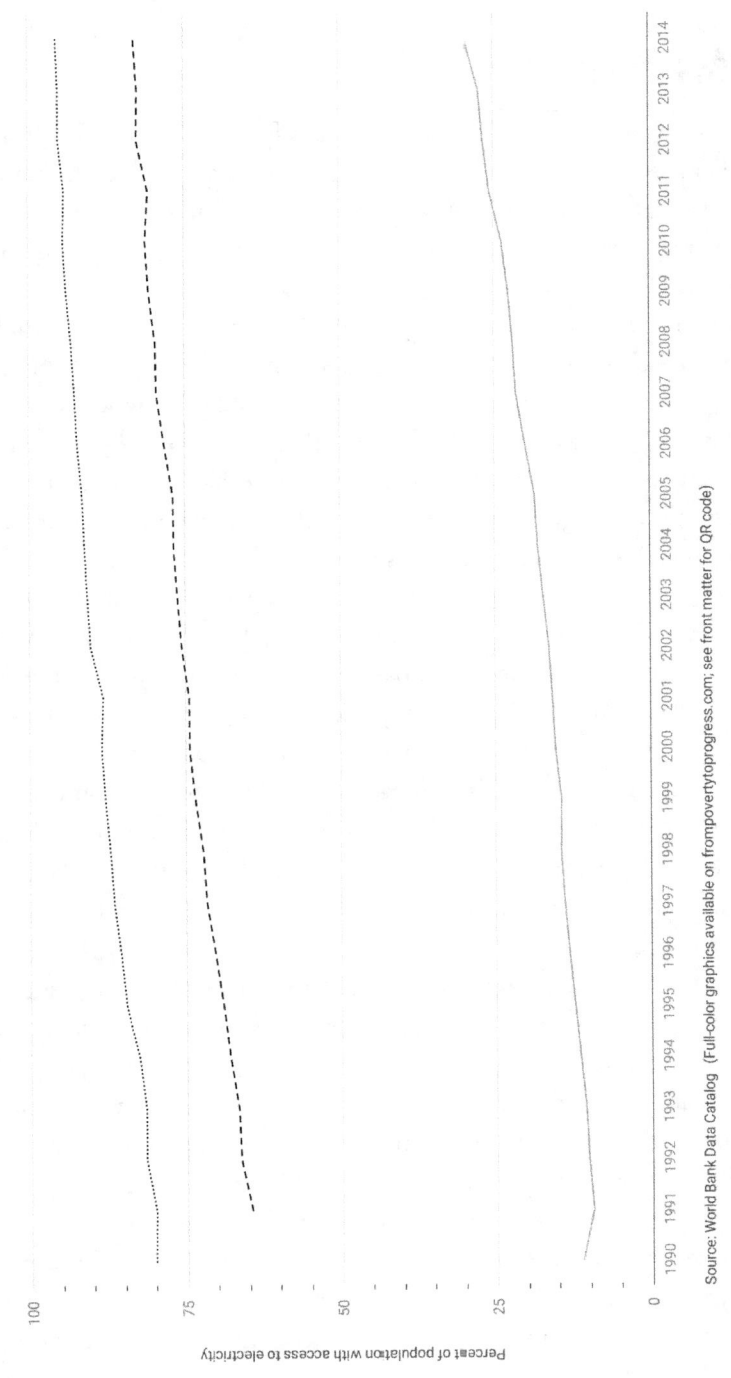

Percent of population with access to electricity

1990 1991 1992 1993 1994 1995 1996 1997 1998 1999 2000 2001 2002 2003 2004 2005 2006 2007 2008 2009 2010 2011 2012 2013 2014

Source: World Bank Data Catalog (Full-color graphics available on frompovertytoprogress.com; see front matter for QR code)

Populous 12 — Bottom 20 ∙∙ Transformative 16

Life Expectancy

Living a long life is an important factor for most of us. We all would prefer a long life to a short life. With improved access to drinking water and sanitation, two of the key causes of early death are being removed. This improvement has affected life expectancy at birth for women. For this metric, we have data from 1960 to 2016.

For the Wealthy 12, every nation shows strong improvement in life expectancy. The average increased from 74 years in 1960 to 84 years in 2016. In 1960, every nation had a life expectancy for women below 76 years old. By 2016, every nation in the Wealthy 12 had a life expectancy for women at 81 or better. Given that there is likely a finite limit to how long humans can live, it is not clear whether this trend will continue, but up to now the positive trend has been clear.

For the Populous 12, the trend is even stronger. In 1960, every nation had a life expectancy below 60 years of age for women. By 2016, all but Congo and Nigeria surpassed this level; however, both of these nations still showed substantial progress. The average in 1960 was 47 years, while it has increased to 71 years in 2015. Not a single nation within the Populous 12 lacked progress in life expectancy, and all trend lines point to continued progress. In addition, four nations — Mexico, Turkey, China and Iran — are nearing the levels of the Wealthy 12.

The trend is similar for the Bottom 20. In 1960, all nations had a life expectancy below 50 years of age for women and many of them were far below that level. By 2016, every single nation had surpassed that level. The average increased from 39.6 years in 1960 to 62.5 years in 2015. Even Rwanda, which experienced a devastating genocide in 1994, managed to reach 69 years by 2016. As with the other nations being studied, not one nation trended downward, with the temporary exceptions of civil-war-torn Rwanda and Sierra Leone.

In 1960 the life expectancy of women at birth varied greatly between nations within the Transformative 16. Some of the nations (Israel, Ireland, Spain, Japan and Hong Kong) had already reached the levels of the Wealthy 12. Some, such as China, India and Indonesia,

Life Expectancy at Birth for Women

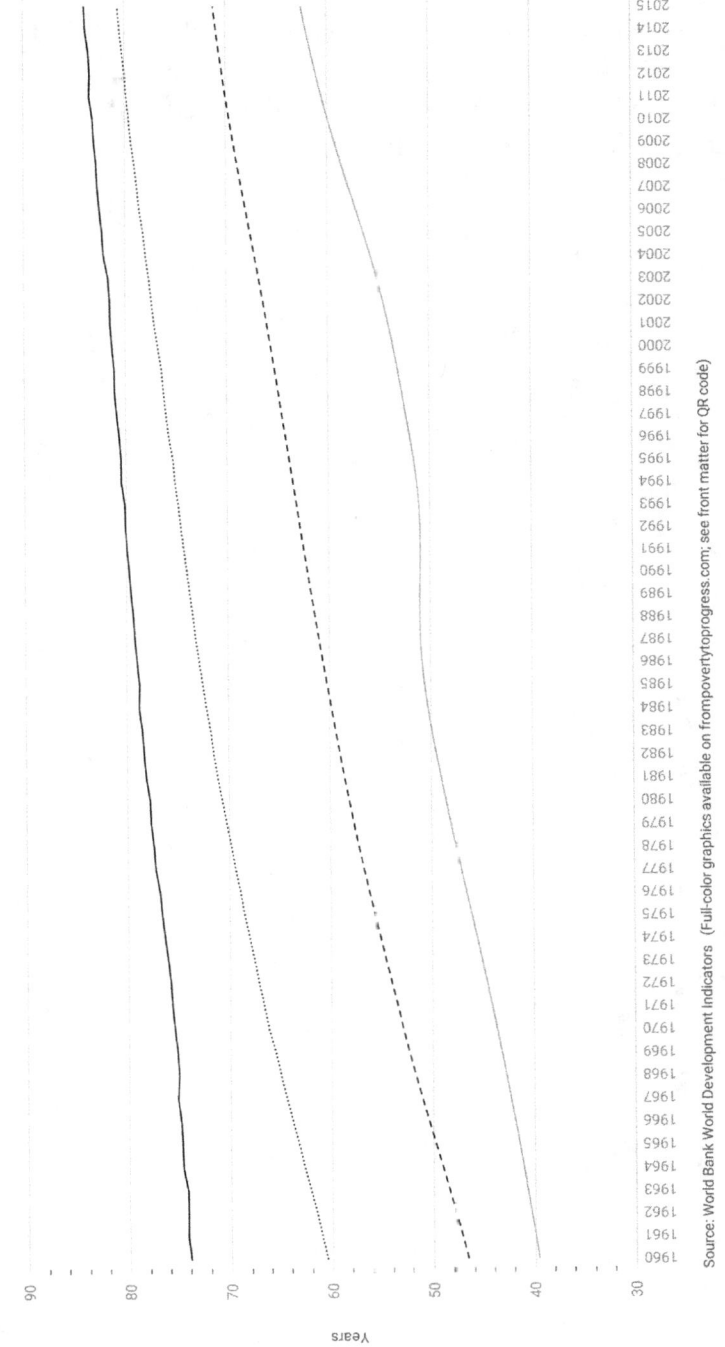

Years

90
80
70
60
50
40
30

1960 1961 1962 1963 1964 1965 1966 1967 1968 1969 1970 1971 1972 1973 1974 1975 1976 1977 1978 1979 1980 1981 1982 1983 1984 1985 1986 1987 1988 1989 1990 1991 1992 1993 1994 1995 1996 1997 1998 1999 2000 2001 2002 2003 2004 2005 2006 2007 2008 2009 2010 2011 2012 2013 2014 2015

Source: World Bank World Development Indicators (Full-color graphics available on frompovertytoprogress.com; see front matter for QR code)

— Wealthy 12   ▪ ▪ Populous 12   ⁓ Bottom 20   ▪▪ Transformative 16 lagged far behind. By 2016 all of the nations in this group had reached levels of 70 years or higher. Virtually all of them had exceeded the life expectancy levels that the Wealthy 12 had in 1960, and many of them exceeded the levels of the Wealthy 12 today.

Neonatal Mortality

Neonatal mortality rates per 1,000 live births from 1960 to 2017 show very similar trends, but they are even bigger in magnitude. Neonatal mortality rates measure the proportion of children who die within the first 28 days of their life. This terribly sad event is becoming increasingly uncommon.

For the Wealthy 12, every nation in 1960 had a neonatal mortality rate above 12 deaths per 1000 births. In 1960 the average was 14.9 deaths. Just think of that. In the richest nations of the world, within the lifetime of many people today, one in every 71 children died within the first 28 days of their lives.

By 2017 every nation within the Wealthy 12 had managed to push that level below 4 deaths, and the average was 2.5 deaths. While the improvements are slowing down, this is likely because the rates are nearing zero.

For the Populous 12, every nation had a neonatal mortality rate above 50 deaths in 1960 and the average was 65. As a parent, it is hard for me to imagine the emotional hardship that this must have caused parents at that time. By 2019, every nation was below this level; the average had dropped to 20 deaths and for most nations it was below 15 deaths. Not one nation showed anything other than a precipitous decline in neonatal mortality.

A very similar trend has taken place for the Bottom 20. While every nation started above 50% in the 1960s, every nation was below 42% in 2017. The average declined from 61 deaths to 25 deaths. In 2017 about half of these nations were at or near the levels that the Wealthy 12 had been at in 1960.

The Transformative 16 also drove down neonatal mortality to levels below the level of the Wealthy 12 in 1960. The average for the Transformative 16 in 1960 was 40 deaths, while it had dropped to 6 deaths in 2019. In addition, many of these nations have lower levels than the Wealthy 12 do today. The only negative trend has been in Botswana where the AIDs epidemic pushed up the rate of neonatal mortality after 2005. Fortunately, there is hope that this is just a temporary aberration.

Neonatal Mortality

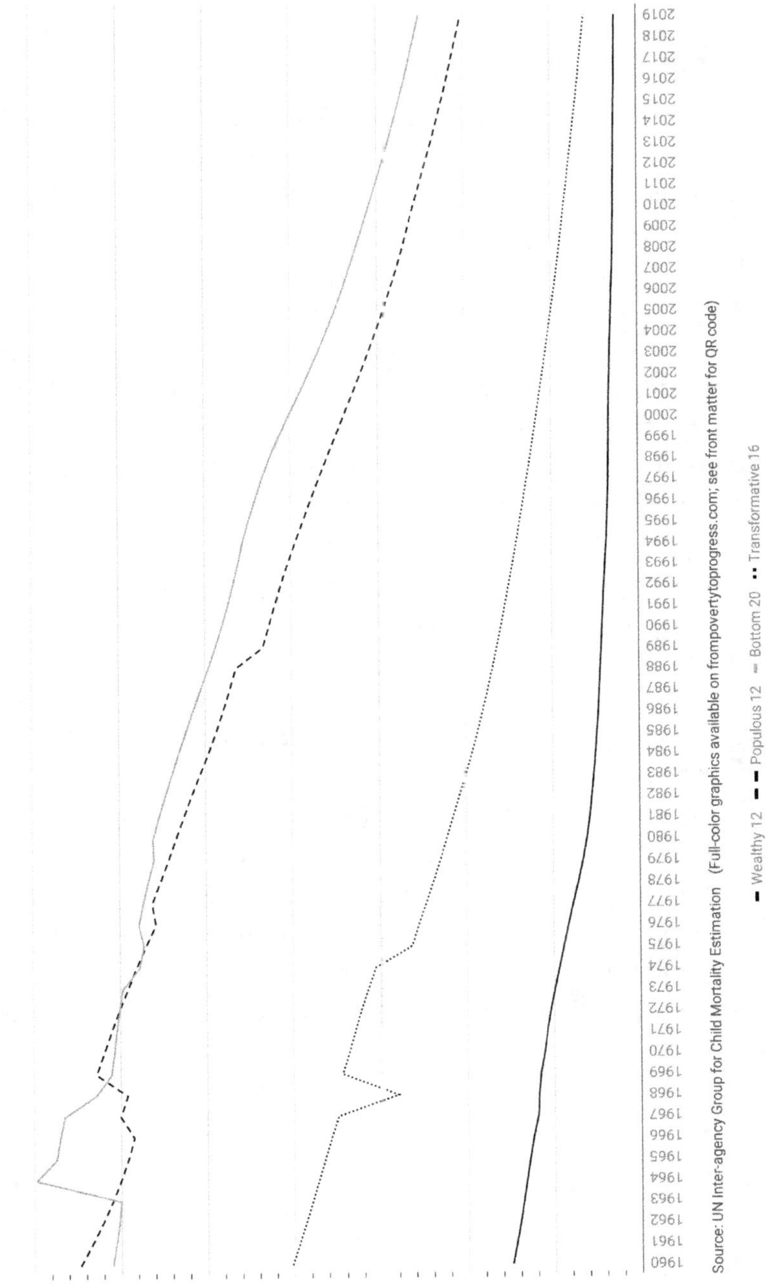

Neonates dying before reaching 28 days of age (per 1,000 live births)

Source: UN Inter-agency Group for Child Mortality Estimation   (Full-color graphics available on frompovertytoprogress.com; see front matter for QR code)

— Wealthy 12   ▬ ▬ Populous 12   ▬ Bottom 20   ••• Transformative 16

Tropical Diseases and Malaria

One of the great scourges of humanity in tropical regions has been disease, particularly malaria, schistosomiasis, yellow fever, sleeping sickness and dengue. Not only have these diseases killed large swathes of society, but they have also sapped the energy of those who survived. These diseases limited development in entire regions.

While people use the term "tropical diseases" today, it is important to understand that many of these diseases were once prevalent outside of tropical regions, although they were less of a problem. Malaria was once endemic in Europe. The disease reached its heights in northern Europe during the late 15th and early 16th Centuries. The disease persisted in Mediterranean Europe as late as the 1970s.

Fortunately, humanity appears to have reached a fundamental turning point regarding its vulnerability to tropical diseases and malaria. The world average rate of infection has been cut in half in less than 30 years, dropping from just under 30,000 in 1990 to 16,290 in 2017. For the Bottom 20 nations, the rate has dropped from 61,000 to just fewer than 40,000. Progress since 2006 has been particularly encouraging.

Every nation within the Wealthy 12, Populous 12, Bottom 20 and Transformative 16, except Niger, either had very few deaths from these diseases in 1990 or experienced strong progress from that date. The large drops in these diseases since 2006 in tropical Africa are particularly promising. The eradication of these terrible diseases is no longer a naïve fantasy; it is happening right before our eyes.

MEASURING PROGRESS

Tropical Diseases and Malaria

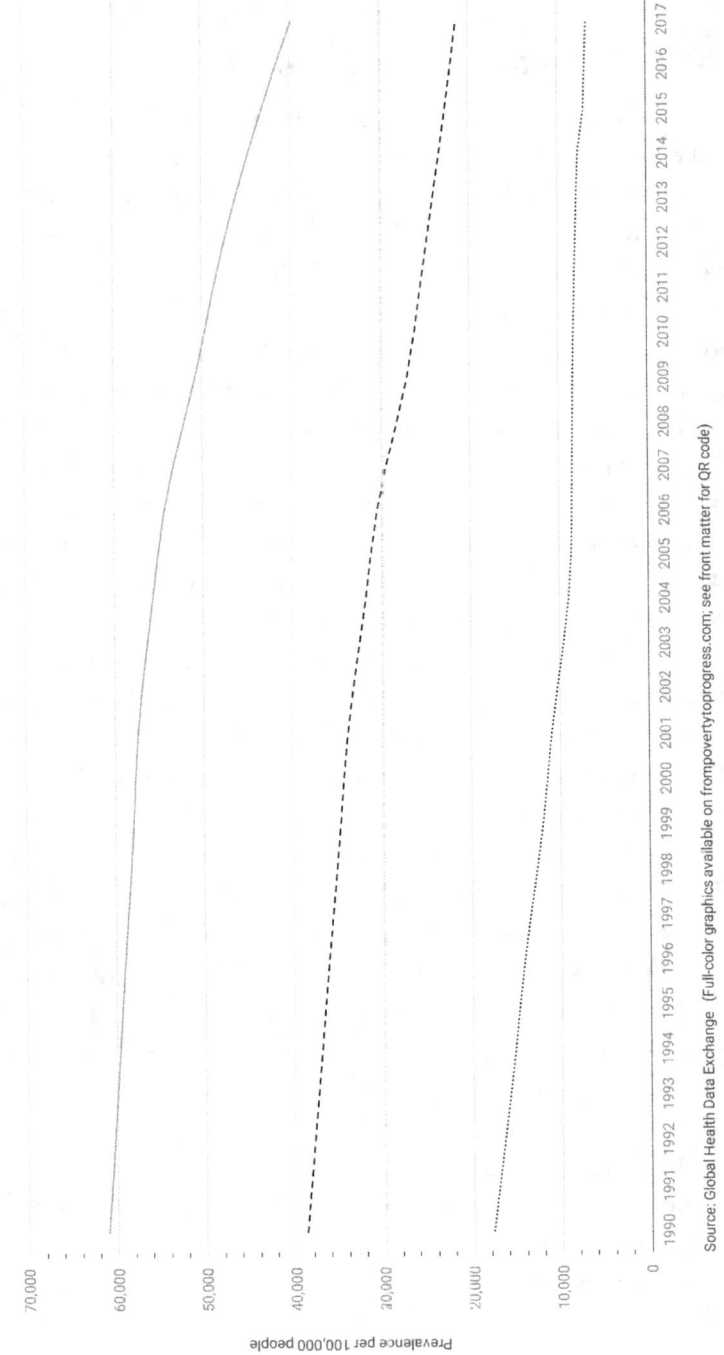

Prevalence per 100,000 people

70,000   60,000   50,000   40,000   30,000   20,000   10,000   0

1990 1991 1992 1993 1994 1995 1996 1997 1998 1999 2000 2001 2002 2003 2004 2005 2006 2007 2008 2009 2010 2011 2012 2013 2014 2015 2016 2017

▬ ▬ Populous 12   ▬ Bottom 20   ▪▪ Transformative 16

Source: Global Health Data Exchange   (Full-color graphics available on frompovertytoprogress.com; see front matter for QR code)

Urbanization and Slum Dwellers

One fairly well known trend has been increased urbanization throughout the entire world. This trend is found in all four of our categories.

In 1960 the Wealthy 12 nations had an urbanization rate of 71%. By 2019 they had reached a level of 84%. Every nation within the group increased its urbanization rate during this period. Between 1960 and 2019 the Populous 12 nations experienced a profound transformation. On average, the urbanization rate doubled from 26% to 56%. Within the Bottom 20 nations, the trend is the same, although with a lower baseline. The average urbanization rate more than tripled from 10% in 1960 to 34% in 2019. The increase in urbanization rates is also evident within the Transformative 16 nations. The average urbanization rate increased from 44% in 1960 to 74% in 2019. Seven of these nations now have urbanization rates that resemble the Wealthy 12. Only India lags behind on this metric.

One false perception that Westerners have of cities in the developing world is that they are dominated by massive slums. If this were true, then this would negate many of the benefits of urbanization. Fortunately, rapid urbanization is occurring alongside declining numbers of slum dwellers.

Housing quality is an important factor in the standard of living. Unfortunately, there are no good international metrics in this domain. The closest that I could find is the percentage of urban populations that live in slums. We have data for this metric between 1990 and 2014. The United Nations defines a slum household "as a group of individuals living under the same roof lacking one or more of the following conditions:

- Access to improved water
- Access to improved sanitation
- Sufficient living area
- Security of tenure"

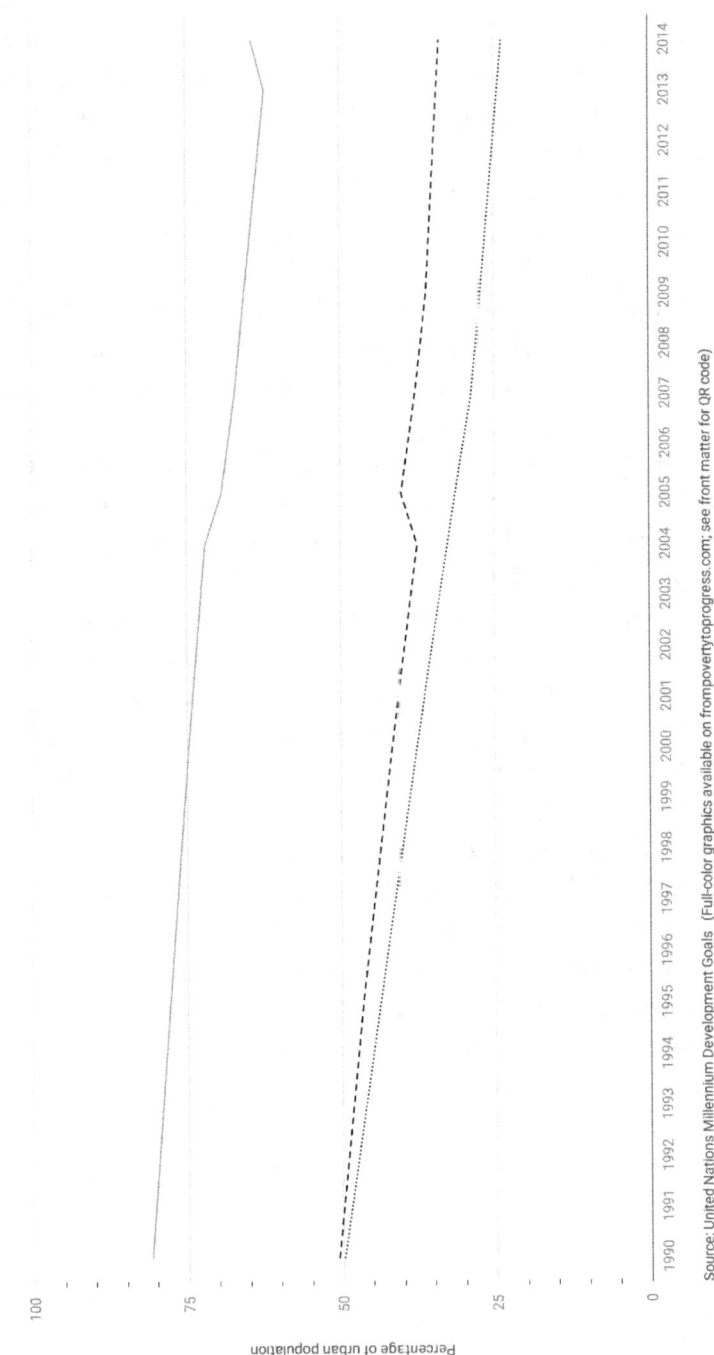

Slum Dwellers

Percentage of urban population

100 · · · · · · · 75 · · · · · · · 50 · · · · · · · 25 · · · · · · · 0

1990 1991 1992 1993 1994 1995 1996 1997 1998 1999 2000 2001 2002 2003 2004 2005 2006 2007 2008 2009 2010 2011 2012 2013 2014

Source: United Nations Millennium Development Goals (Full-color graphics available on frompovertytoprogress.com; see front matter for QR code)

- - Populous 12   -- Bottom 20   ·· Transformative 16

For the Wealthy 12, this metric is useless as all the nations are at or near zero. For the Populous 12, however, there is strong evidence of progress. The percentage of urban dwellers living in slums in the Populous 12 nations dropped from 51% in 1990 to 34% in 2014.

The only nations in the Populous 12 without clear progress are Pakistan and Congo (which only has data going back to 2005). Mexico and Turkey had the lowest rate of slum dwellers in 1990 (both around 23%), but they both managed to cut their levels in half by 2014. Most of the other nations also made significant gains. To do so in the middle of the most rapid expansion of urban population that the world has ever known is particularly impressive.

The trends for slum-dwelling for the Bottom 20 are much more mixed. While the average dropped from 81% in 1990 to 64% in 2014, a few nations did not show progress. Among countries with data going back to 1990, the trends for the Central African Republic, Mozambique, and Malawi are flat or even slightly worse. It is important to note, however, that all of the nations with data in 1990 had levels over 66%, while 10 of the 16 had levels lower than this in 2014.

While many of the nations in the Transformative 16 are missing data, China, India and Indonesia all saw dramatic drops in the percentage of city dwellers living in slums from 1990 to 2014. The highest level in 2014 (China at 25%) was significantly lower than the lowest level in 1990 (China at 44%). Both India and Indonesia managed to cut their percentage of slum dwellers in half.

It seems quite likely that nations with missing data did achieve the same progress in eliminating slums. The fact that this occurred during a period of dramatic increases in the urbanization rate is particularly impressive.

Cereal Production

An efficient food production and distribution system is one of the Five Keys to Progress. Production of cereals is particularly important. For 10,000 years the majority of the calories consumed by humans in agricultural societies consisted of cereals. Wheat, corn and rice are still the dominant forms of human consumption in developing countries. In poor countries, they often make up 80% of the calories consumed.

For this reason, increasing cereal production is critical to progress. We will use the net production of cereals from 1961 to 2016 as the metric for measuring cereal production per capita. The metric compares cereal production per capita in a given year in comparison to cereal production per capita in 2004-2006 in the same nation.

Cereal production in the Wealthy 12 increased from 47 in 1960 to 112 in 2014. Except for the Netherlands, all nations within the Wealthy 12 substantially increased cereal production. Just as importantly, they managed to do so while expending far more effort increasing agricultural production in more high-value crops and animals, such as dairy and meat.

The trends for the Populous 12 are even more positive. While wealthy nations had reached a level of development that enabled a varied diet by 1960, the Populous 12 nations were still struggling with famine and malnutrition. Being unable to afford imported cereals, increasing domestic food production was essential to progress.

While every nation in the Populous 12 had levels of less than 37 in 1961, by 2016 all but Iran had levels of 97 or higher. Ethiopia, famous for its famines during the 1980s, was a star: going from 45 in 1993 to over 200 in 2016. It should not be surprising that famines have disappeared from all of these nations, an extraordinary accomplishment given their history.

The trend of cereal net production in the Bottom 20 is not quite so clear-cut. Before 1990, most of these impoverished nations had only a small upward trend and some were going downhill. Famines were still a very real threat in many of these countries.

Starting around 1990, however, cereal production began to take off. While in 1990, all of the Bottom 20 were below 103 (and most were far below it), by 2016 all but three (Burma, Burundi and Gambia) were above it. The average level for the Bottom 20 rose from 36 to 147. By the 21st Century, the widespread famines of the 1970s and 80s had ceased to haunt the African continent.

Cereal net production within the Transformative 16 also showed some divergence in the outcome. Japan and Taiwan both show a decline in cereal production, while Botswana varies dramatically. Most of the other nations, however, show a clear increase. The average from the Transformative 16 nations increased from 74 in 1961 to 115 in 2014. Note that one of the reasons for the variations in outcome is that many of these nations became rich enough to import cereal rather than producing it domestically.

Cereal production per capita

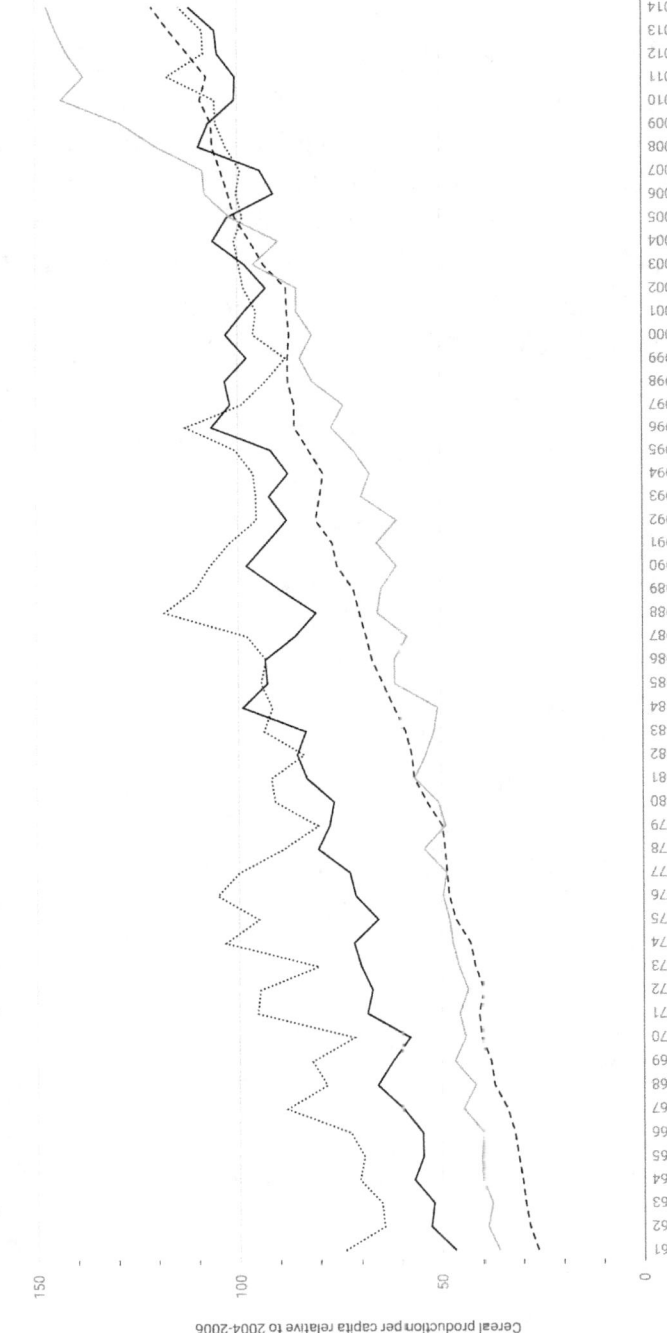

Cereal production per capita relative to 2004-2006

Source: Food and Agriculture Organization of the United Nations) (Full-color graphics available on frompovertytoprogress.com; see front matter for QR code)

Wealthy 12 ‑ ‑ Populous 12 — Bottom 20 ⋯ Transformative 16

World Agricultural Prices

As more and more nations increased their food production, and preservation technologies improved, importing food had become increasingly attractive. Long-term trends in the Grilli-Yang agricultural price index show that prices have dropped substantially since 1920. Given that world populations have grown dramatically since then, this is a truly impressive achievement. Widespread predictions of famines caused by overpopulation have proven spectacularly incorrect.

World Agricultural Prices

Grilli–Yang Index; Price relative to 1977–1979

Source: US Department of Agriculture

Daily Supply of Calories

All of the increased food production and lower food prices have meant that almost every part of the world is eating better than in previous generations. For measuring food consumption, I will use the daily supply of calories per person. This statistic is available from 1961 to 2013.

By 1961, all the Wealthy 12 nations consumed at least 2,750 kCal/day. Given this high level of calories, it is not surprising that improvements were relatively small; the average caloric intake increased from 3,020 to 3418. Every nation except Switzerland, which had the highest level in 1961 increased in this period. In all of these nations, unfortunately, obesity has become an increasing problem.

For the Populous 12, there has been a clear trend towards an increased daily supply of calories per person. In 1961, people in all nations but Turkey consumed 2,300 kCal/day or less. China's situation was particularly disastrous, with a caloric consumption of 1,415 kCal/day.

By 2013, the situation for almost all of the nations in the Populous 12 had dramatically improved. The average increased from 2,027 to 2,873. Now all nations surpass 2,300 kCal/day, and half of them are above 3,000. This gives them levels of food consumption that are very close to the Wealthy 12 in the same year. Only Congo and Pakistan failed to show substantial improvement during this period.

The trend of calorie consumption for the Bottom 20 has been far less encouraging. Between 1961 and 2013 the average caloric intake increased from 1,969 to 2,405. Most of the countries showed strong improvement, but the Central African Republic and Afghanistan showed serious declines, while Uganda, Rwanda, Senegal and Congo's trends were fairly flat.

Overall, in 1961 only two nations exceeded 2,300 kCal/day, while in 2013 all but four did so. So, a lack of calories is still a serious problem for almost all of the Bottom 20; however, the levels are not quite as desperate as they were back in 1961.

The Transformative 16 made clear progress in feeding their people. In 1961, only the citizens of Ireland consumed over 3,000 kCal/day,

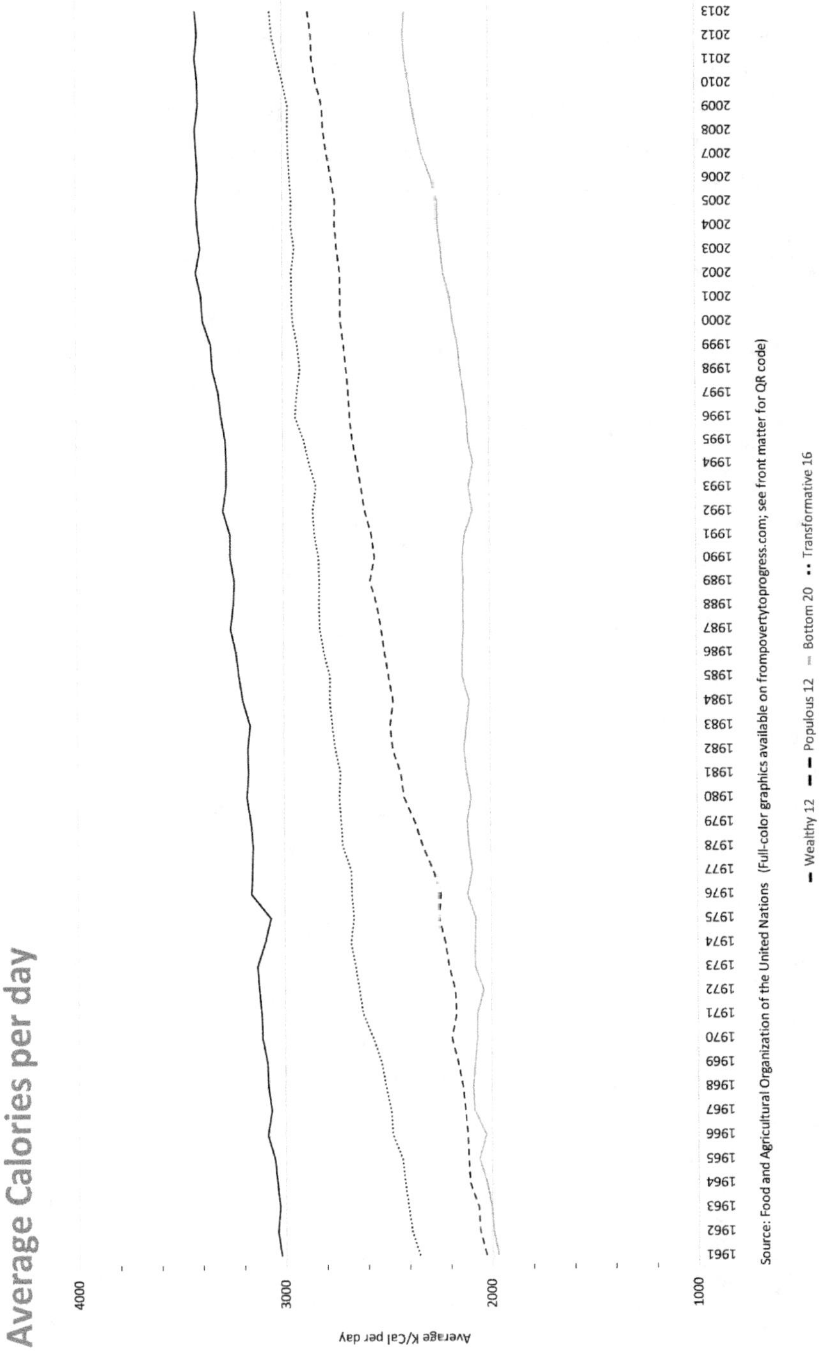

Average Calories per day

Average K/Cal per day

4000

3000

2000

1000

1961 1962 1963 1964 1965 1966 1967 1968 1969 1970 1971 1972 1973 1974 1975 1976 1977 1978 1979 1980 1981 1982 1983 1984 1985 1986 1987 1988 1989 1990 1991 1992 1993 1994 1995 1996 1997 1998 1999 2000 2001 2002 2003 2004 2005 2006 2007 2008 2009 2010 2011 2012 2013

Source: Food and Agricultural Organization of the United Nations (Full-color graphics available on frompovertytoprogress.com; see front matter for QR code)

— Wealthy 12   ▬ ▬ Populous 12   — Bottom 20   •• Transformative 16 while second-place Israel consumed 2,805. Most nations within the Transformative 16 consumed less than 2,500 kCal/day.

From 1961 to 2013 the average caloric intake increased from 2,351 to 3,058. By 2017 every nation in the group exceeded 2,500, and most exceeded 3000 kCal/day, a level on par with the Wealthy 12. Only Japan experienced a decline, and it is not clear that this is hurting their health in any way.

Famines

The net result of higher food production and higher caloric consumption has been the virtual elimination of famines. While the world population has increased five-fold since the 1860s, the number of deaths due to famine has declined dramatically.

It is also important to note that the nature of famines has changed as well. Before 1920, famines were generally drought-induced shortages that could not be alleviated due to poor local transportation systems. Neighboring regions often had a food surplus, but the lack of transportation technologies made it impossible for the food to be transported. Famines of that type were beyond human solutions given the local technology.

Since 1920, the vast majority of famines have been at least partly deliberately induced by Communist regimes as a political weapon against perceived opposition. The Soviet Union, China, Cambodia, Ethiopia, and North Korea all engineered famines that resulted in millions of deaths.

Since 1980 there have been very few deaths due to famine. The exceptions — Congo, Sudan, and Somalia — were mainly due to ongoing civil wars disrupting food production and distribution networks. Thankfully, such disruptions are becoming increasingly rare. The threat of famine that haunted so many generations in the past may completely disappear soon.

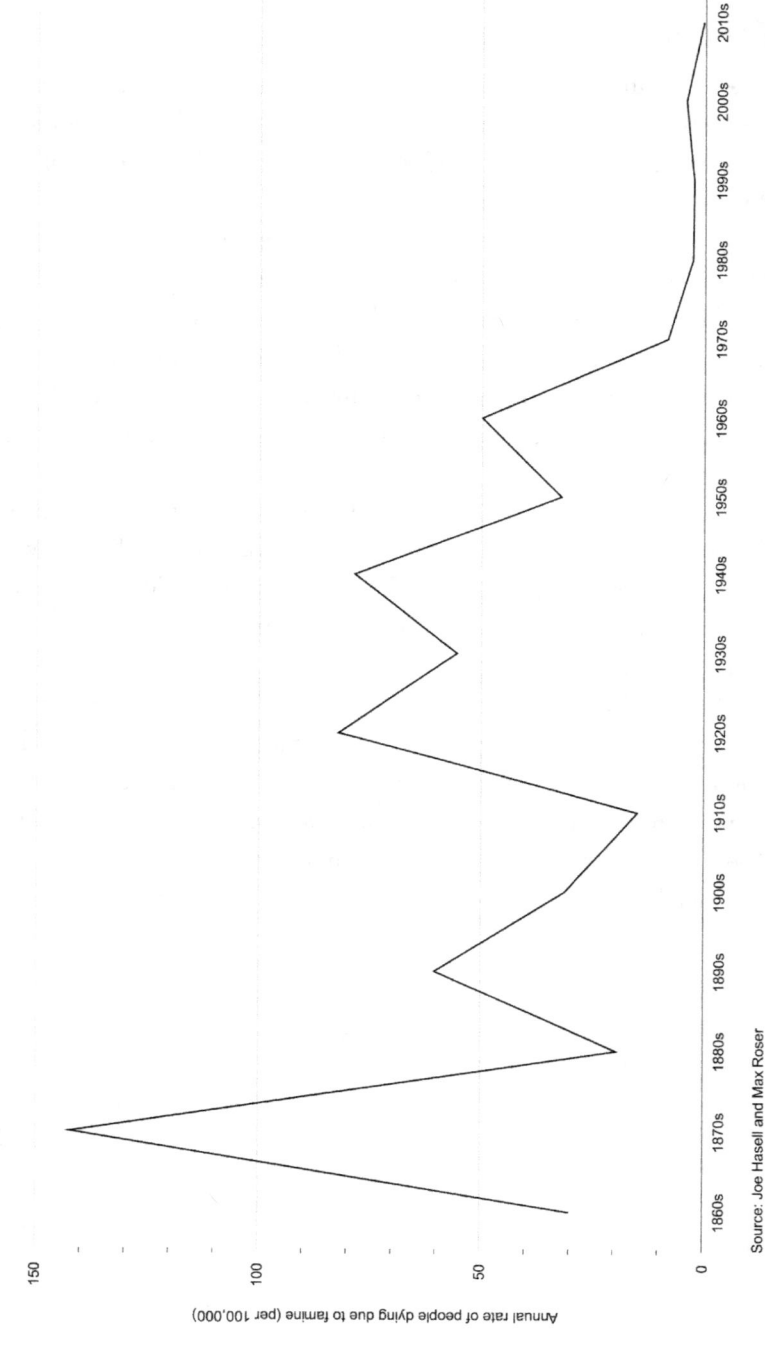

Deaths Due to Famine

Annual rate of people dying due to famine (per 100,000)

Source: Joe Hasell and Max Roser

Education

Education is critically important for humans to experience a well-rounded life. It is also important in transferring the skills necessary for building and using technologies. Education also enables adults to pass on their values to children.

Fortunately, we have a many good metrics measuring the number of years of schooling per young person, so we can make some strong conclusions. The metric of mean years of schooling is particularly useful because, at least for Western nations, the data goes back to 1870.

Since 1870, there has been a tremendous increase in the mean years of schooling in every nation in the Wealthy 12. In 1870, the highest value was Norway, coming in at 4.3 years. By comparison, in Germany it was 2.2, in the UK it was 0.9 and in France it was 0.7 years schooling! Undoubtedly, there were some well-educated children, but a complete lack of schooling for most people was the norm in 1870.

Since 1870, the Wealthy 12 nations have increased the average years of schooling five-fold from 2.5 to 12.8 years. By 2017, every nation in the Wealthy 12 was providing well over 10 years of schooling and most had levels of 11 years or higher.

For the Populous 12, the trend in the number of years of schooling is the same, although progress did not start until the 1960s. As late as 1900, every nation within the Populous 12 provided less than one year of schooling. Since 1870 the average years of schooling have skyrocketed from 0.05 to 7.00 years.

By 2017, all but Ethiopia and Pakistan were at six years or higher. Primary schooling has become universal in almost all of these countries. The levels of education are still much lower than in the Wealthy 12, but the trend is a steep upward curve and there is no reason to believe that it will plateau.

For the Bottom 20, there has been similar progress in the number of years schooling, although the take-off did not occur until the 1980s. While many children in the Bottom 20 are not completing primary school, before 1980 almost no children did. And based upon the steep trend lines, there is every reason to believe that this progress will continue for the foreseeable future.

The Transformative 16 all experienced dramatic increases in mean years of schooling. In 1870, all but Spain had a mean of one year or less, and Spain was not much above that. The take-off occurred at very different times, but each nation in the Transformative 16 increased schooling by an order of magnitude from 0.22 to 9.94 years.

By 2017, seven of the nations had reached the same level of schooling as the Wealthy 12, while the others (China, India, Indonesia and Thailand) show every indication of reaching those levels soon.

Years of Schooling

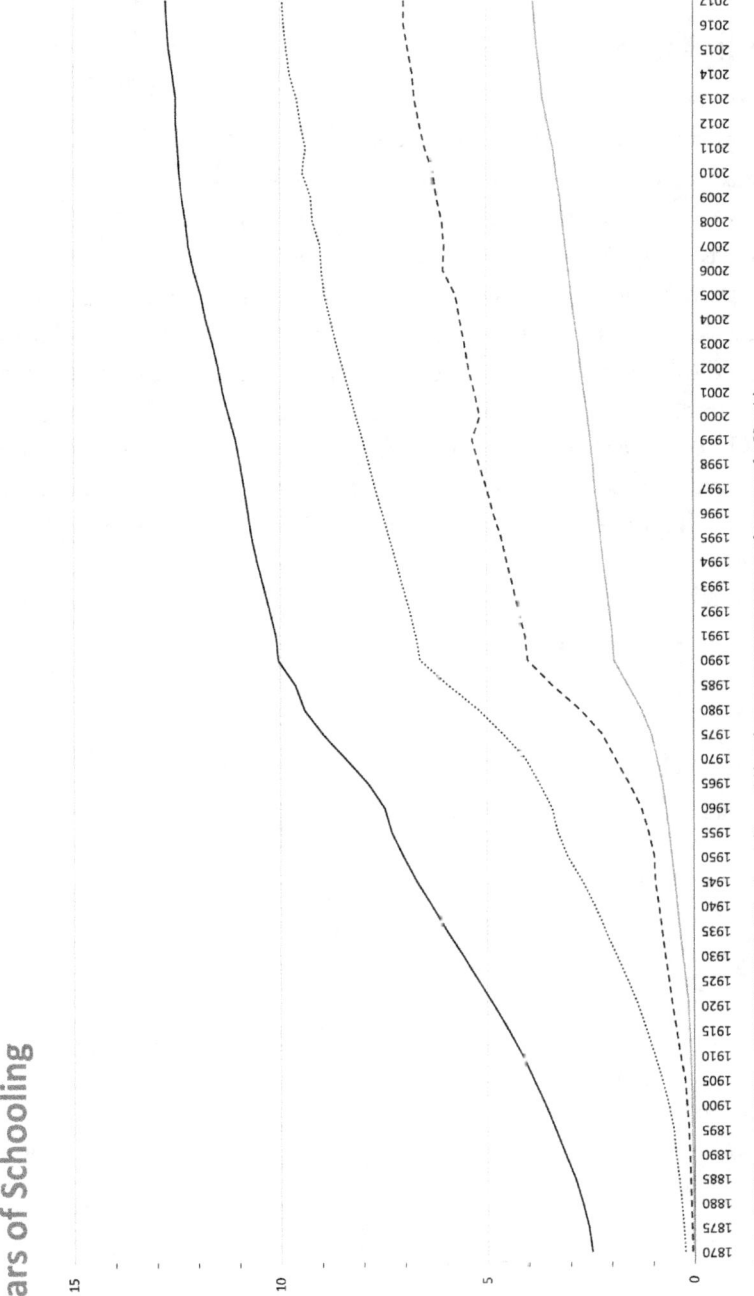

Average total years schooling for adult population

Lee-Lee (2016); Barro-Lee (2018) and UNDP HDR (2018) (Full-color graphics available on frompovertytoprogress.com; see front matter for QR code)

— Wealthy 12   ▪ ▪ Populous 12   ⋯ Bottom 20   •• Transformative 16

Intelligence

While most people think of intelligence as a constant over time, there is a great deal of evidence to the contrary. While the exact reasons for this change are controversial, overall intelligence as measured by IQ tests has increased

by 20 points since the late 1930s. For instance, average intelligence in Europe increased by 23 points, in America by 32 points, while Africa increased by about 14 points. Moreover, these changes are documented on every continent, with Asia showing the largest gains (+37 points).

While the causes and significance of this increased intelligence are unclear, this is undoubtedly a sign of progress. This phenomenon is called the "Flynn effect" for the researcher who first discovered it: James R. Flynn.

Average Age of Women at First Marriage

Girls in agricultural societies typically married very young, often to significantly older men. This meant that women of the past have had little choice but to devote their (short) adult life to child rearing.

As women throughout the world enjoyed the benefit of prosperity, opportunity and freedom, many are choosing to delay marriage. Some women have focused their time on education; some have focused on employment in the workforce; others chose to focus on family and children. Up until quite recently, virtually no women had a choice at all.

For the Wealthy 12, the average age of women at first marriage in 1970 was less than 23 in all nations except Sweden, which was only slightly above that. Some of the data for after 2010 is not available and there is an obvious dip during the 2008 recession, but by 2014, marrying over 30 years old had become the norm. This has given women greater opportunities to invest in their education and careers than in previous generations.

For the Populous 12, the trend has not been quite so evident. The average age at first marriage was 20 in 1978. By 2014, their age had increased to 23 years old. While in 1970, the age of women at first marriage was 21 or lower for all nations, except Brazil, by 2014 every nation had increased in age.

Among most nations within the Bottom 20, the trend is clear but not strong in magnitude. In the 1960s, all nations with data had ages of 19 or less. By 2014, the majority were over 20 years old. The average age increased from 19 to 21 years old. These are not huge differences, but it seems to be clear at least that marriage under the age of 18 has declined substantially in the poorest nations of the world.

Within the Transformative 16, all nations show a trend of women marrying later. The average age increased from 23 to 26 years old. By 2012 many reached ages similar to the Wealthy 12, while in other countries teenage brides are becoming a thing of the past.

Age of Women at First Marriage

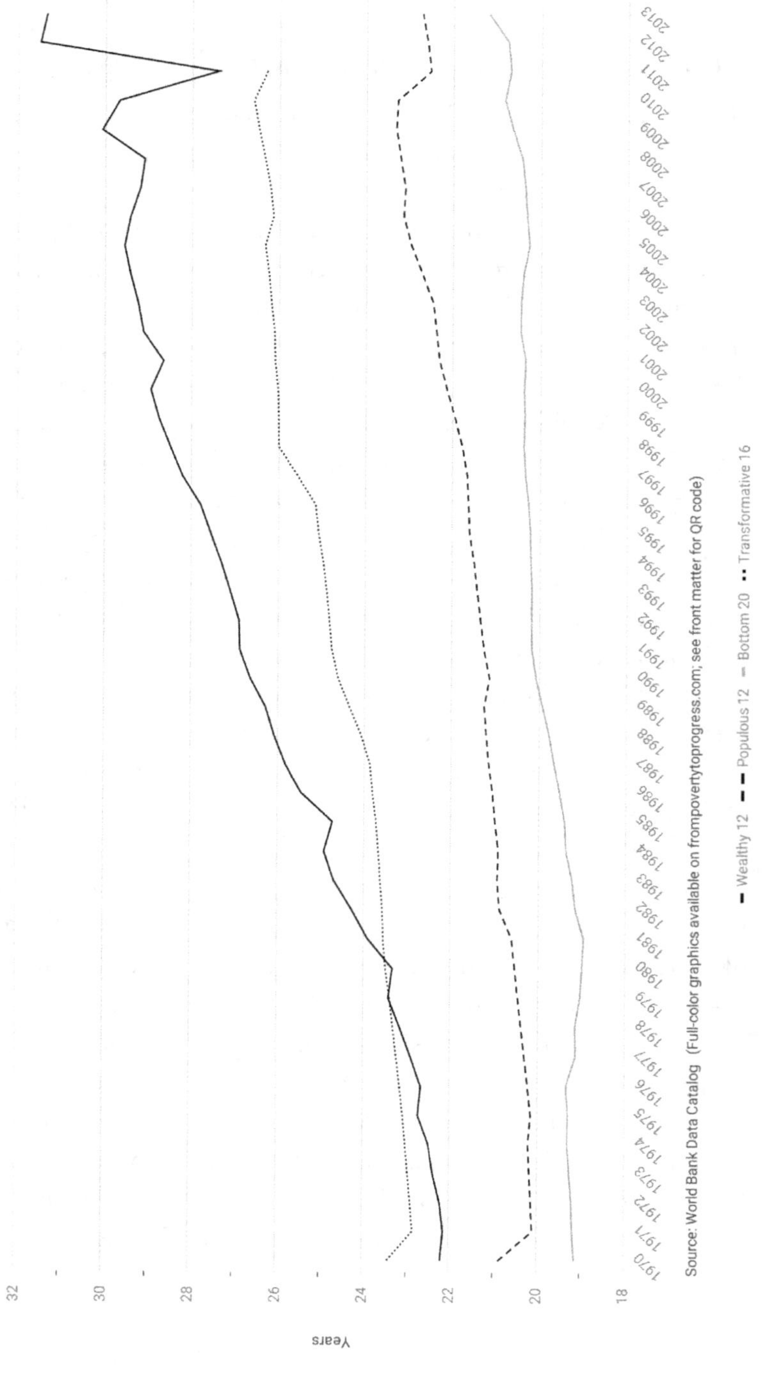

Years

32   30   28   26   24   22   20   18

1970 1971 1972 1973 1974 1975 1976 1977 1978 1979 1980 1981 1982 1983 1984 1985 1986 1987 1988 1989 1990 1991 1992 1993 1994 1995 1996 1997 1998 1999 2000 2001 2002 2003 2004 2005 2006 2007 2008 2009 2010 2011 2012 2013

Source: World Bank Data Catalog (Full-color graphics available on frompovertyprogress.com; see front matter for QR code)

— Wealthy 12   - - Populous 12   — Bottom 20   ⋯ Transformative 16

Democracy

While we take democracy for granted today, it is important to realize that, for virtually all of human history, it was a rarity. Monarchs, chiefs or clan leaders ran the vast majority of societies in the past.

In the early 19th Century, the United States was probably the only nation in the world to qualify as a democracy. Even as late as 1945, democracy was restricted to Northwest Europe, the United States, Canada, Australia and New Zealand. Since that time, there has been a gradual spreading of democratic governance until it encompassed the majority of the world's population.

Economic Freedom

An important cause of economic growth has been the government giving people the freedom to solve each other's problems. The Economic Freedom Index gives us a means by which to measure this freedom. The index ranges from 0 (no economic freedom) to 10 (maximum possible economic freedom). We have data for this metric from 1970 to 2016.

The Wealthy 12 started out with moderately high levels of economic freedom in the 1970s with Canada and the United States being the freest and Sweden being the least free. During the 1980s there was a clear movement towards greater economic freedom, and then relative stability at high levels of freedom after 2000.

The average level of economic freedom increased from 6.4 in 1975 to 7.9 in 2015. By 2016, France had the lowest level of economic freedom (7.25), which was higher than the majority of the Wealthy 12 in 1970. Not one member of the Wealthy 12 failed to experience a rise in economic freedom during this time period.

The Populous 12 experienced much the same improvement in economic freedom, except those nations started from much lower levels. The average level of economic freedom increased from 4.2 in 1980 to 6.4 in 2015.

From 1970 to the mid-80s, the trend was generally negative, but the 1980s appeared to be a turning point for most nations in the Populous 12. By 2016, all the nations within the Populous 12 had levels of economic freedom above 5; only Iran saw a decline from their 1970 level.

The Bottom 20 saw a similar trend in increasing levels of economic freedom, except the turning point for them appears to be the early 1990s. The average level of economic freedom increased from 4.2 in 1984 to 6.1 in 2015.

Just as with the Populous 12, there was a clear downward trend from the 1970s that lasted through the 1980s. But then after 1990, a sudden reverse took place. Whereas almost all the Bottom 20 was below 5 in 1990, they were all above 5 in 2016.

Economic Freedom Index

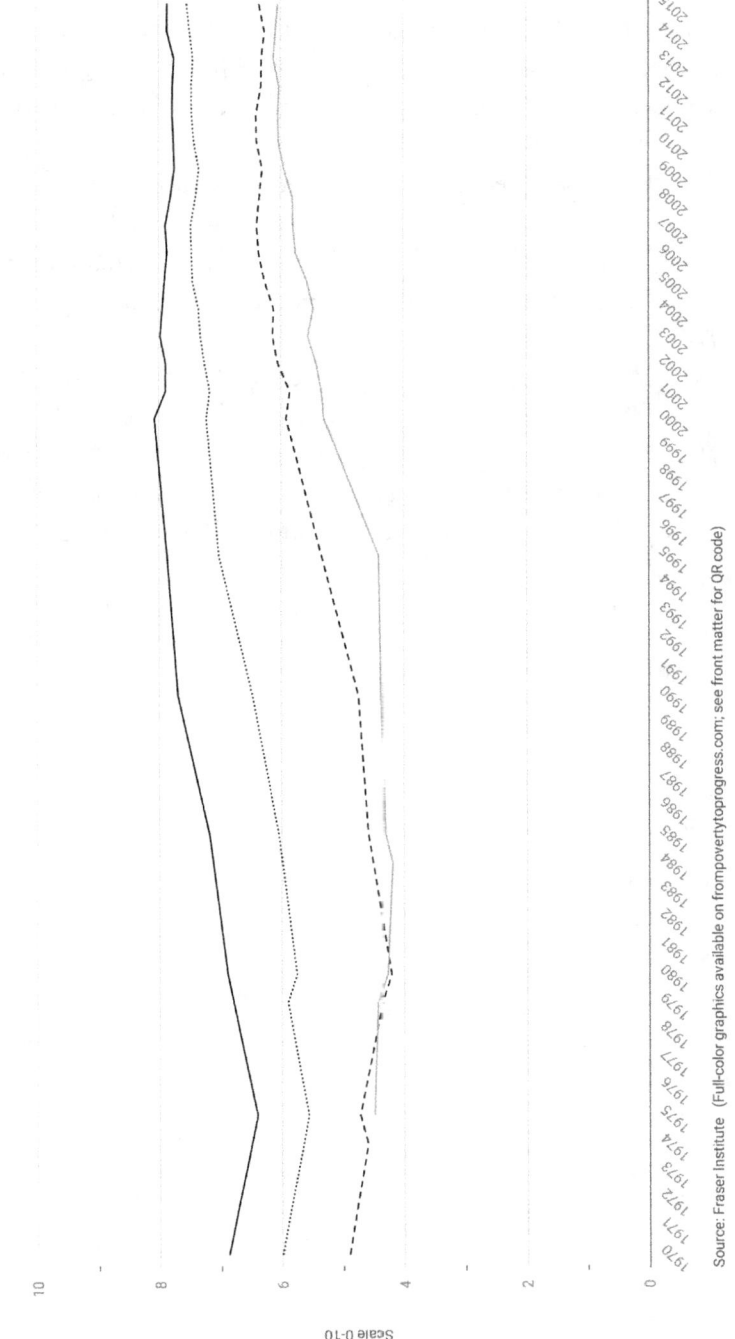

Scale 0-10

Source: Fraser Institute (Full-color graphics available on frompovertytoprogress.com; see front matter for QR code)

— Wealthy 12 - - Populous 12 ~ Bottom 20 ⋯ Transformative 16

Some of the increases in economic freedom among the Bottom 20 were fairly dramatic. Rwanda, Gambia, Ghana and Tanzania all reached levels of economic freedom similar to the Wealthy 12. Burma also experienced a fairly dramatic transformation.

The Transformative 16 all experienced substantial growth in economic freedom. Indeed, many authors claim that it was this increase in economic freedom that enabled their economic transformations in the first place. The average level of economic freedom increased from 5.6 in 1975 to 7.5 in 2015.

While Japan, Singapore and Hong Kong all had high levels of economic freedom in 1970, the other nations varied greatly. By 2016, the lowest level (China at 6.46) was higher than two-thirds of the nations in 1970. The only nation that declined in levels of economic freedom was Japan, which started at a high level and declined only slightly.

Slavery and Forced Labor

KEY INSIGHTS

The broad-based improvement in so many metrics cannot possibly be explained away by increased inequality.

Today, many Westerners view slavery as something unique to the United States and European empires in the New World. In fact, slavery has been a common practice in most pre-industrial societies. While it is impossible to get solid data on the proportion of humans who were slaves at any one time, it is possible to look at the number of countries and colonies where slavery was legal.

In 1800, about 60 countries and colonies had legal slavery. Since then, there has been a steady long-term decline to single digits in 1970. Today, thankfully, legal slavery has all been eliminated from the planet.

Though we do not have firm data to back up overall impressions, it is clear that all types of forced labor (serfdom, peonage, debt bondage, forced labor camps, indentured servitude or the corvee) and other types of less than completely free labor, such as sharecropping have declined dramatically or disappeared completely.

Today the world is dominated by free labor. People are allowed to choose where they want to work, and when they want to quit, and they get rewarded at least somewhat in proportion to their contributions towards producing goods or services.

When one thinks of the enormous physical hardships and emotional deadening caused by hundreds of millions of people forced to work without compensation, this is an enormous accomplishment that should not be ignored.

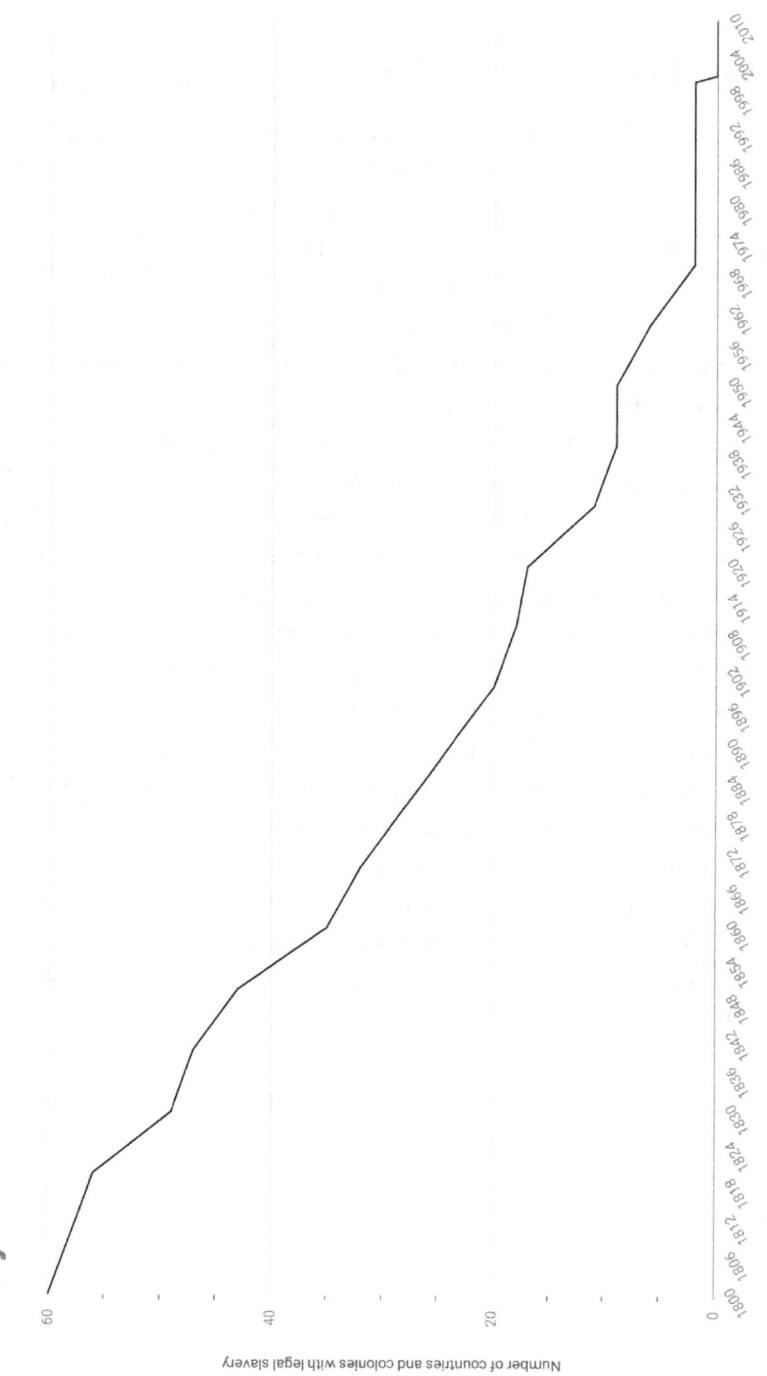

Slavery

Number of countries and colonies with legal slavery

60

40

20

0

1800 1806 1812 1818 1824 1830 1836 1842 1848 1854 1860 1866 1872 1878 1884 1890 1896 1902 1908 1914 1920 1926 1932 1938 1944 1950 1956 1962 1968 1974 1980 1986 1992 1998 2004 2010

Source: Abolition of Slavery timeline Wikipedia

War

Until recently Great Powers were routinely at war. While today we see peace as the norm and war as the aberration, it was not too long ago that the opposite was true.

Armed conflict between Great Powers was the norm from 1500, the time when the modern state began to evolve in Europe, until the Industrial Revolution. The United Kingdom and France were in near-perpetual conflict for centuries. Spain, Prussia, Austria and Russia were also involved in conflicts with each other for a significant period of time.

After the end of the Napoleonic Wars and the Industrial Revolution, a new period of peace between the Great Powers broke out. That peace between the Great Powers lasted for one century. From 1815 to 1913, there were only a few short conflicts between the Great Powers.

World War I and World War II, perhaps the two greatest armed conflicts in global history, abruptly brought the Great Peace to an end. The early parts of the Cold War suggested that the 19th Century Peace had been a short-term phenomenon, but surprisingly the United States and the Soviet Union never directly went to war.

While it would be optimistic to predict that Great Power conflicts have ended forever, the long-term peace since 1945 has now lasted for 75 years with no indications of it ending. It is now World War I and World War II that look to be deviations from the long-term trend. Great Powers are clearly far more interested in generating long-term economic growth than fighting wars. This shift has brought an enormous peace dividend to all of humanity.

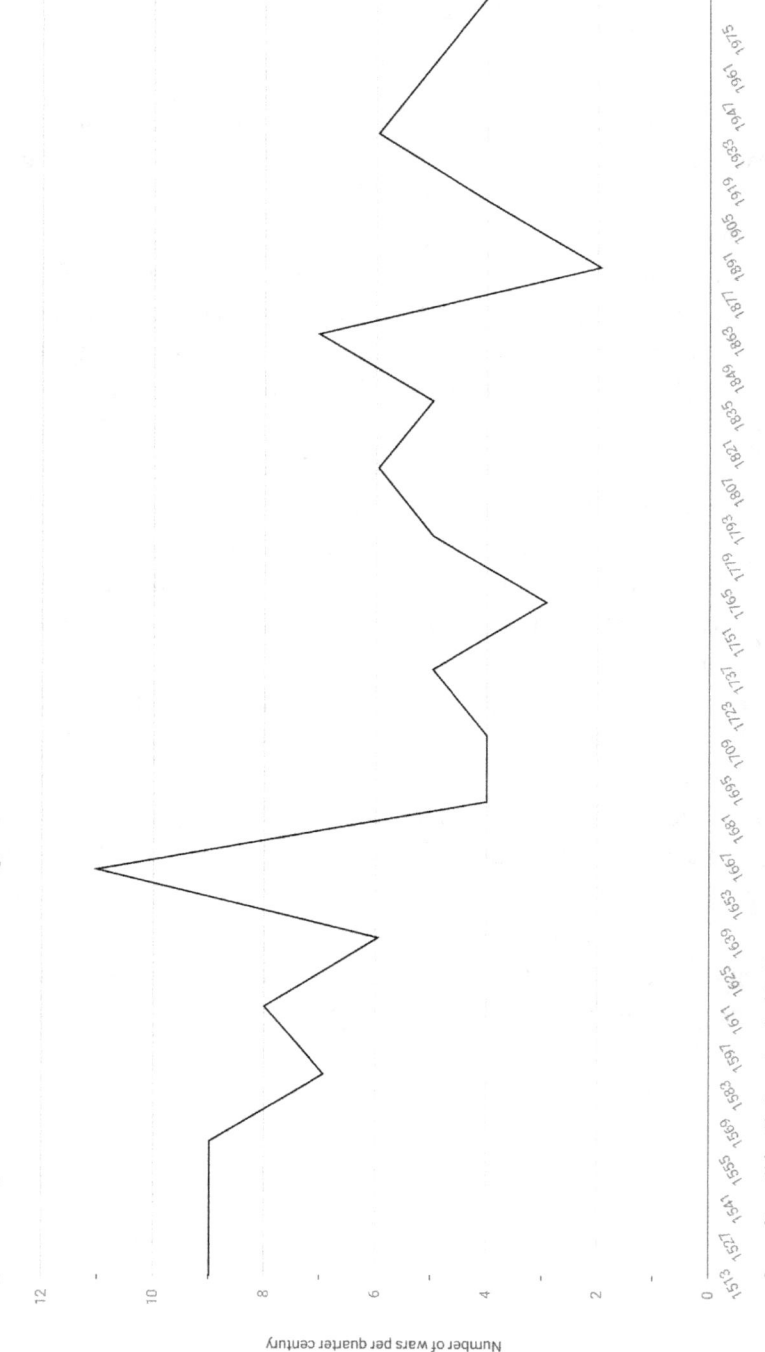

Frequency of Wars Involving the Great Powers

Number of wars per quarter century

Source: Steven Pinker. The Better Angels of Our Nature

Deaths in Wars

As a result of the decline in the number of wars between Great Powers, there has been a general decline in the number of deaths caused by either interstate or civil wars. In 1950 about 76,000 battle deaths occurred per year per conflict. By the year 2000, the rate had dropped to 6,700 — less than one-tenth of the previous level.

While there was no general war between the United States and the Soviet Union, there were a number of smaller wars following World War II. Most of these post-World War II conflicts involved one of two great causes.

The first cause was the desire of African and Asian societies to win their independence from the European empires, particularly Britain and France. The second cause was Communist insurgencies that were funded by the Soviet Union, China and other Communist regimes.

The colonial wars for independence came to an end by 1975, when the last of the European empires — the Portuguese — finally collapsed. Unfortunately, this is also when many of the Communist insurgencies were at their height. However, the gradual withdrawal of funding to insurgencies under Gorbachev and the final collapse of the Soviet Union subsequently eliminated most Communist insurgencies as well.

The years since 2000 have certainly have experienced conflict: Iraq, Syria, Afghanistan, Mexico, Sudan and Yemen have all had significant conflicts with over 10,000 deaths per year in at least one of the last 20 years. It is important to note, however, that three of these conflicts (Afghanistan, Sudan, and Yemen) had their origins during the Cold War.

Since 2014 the number of deaths due to ongoing conflicts has been cut in half — from 235,000 per year to 113,000. As of 2019, all of these ongoing conflicts, except Afghanistan and Mexico, appear to be diminishing in the number of deaths. In addition, many of these conflicts have hope for real peace in the near future.

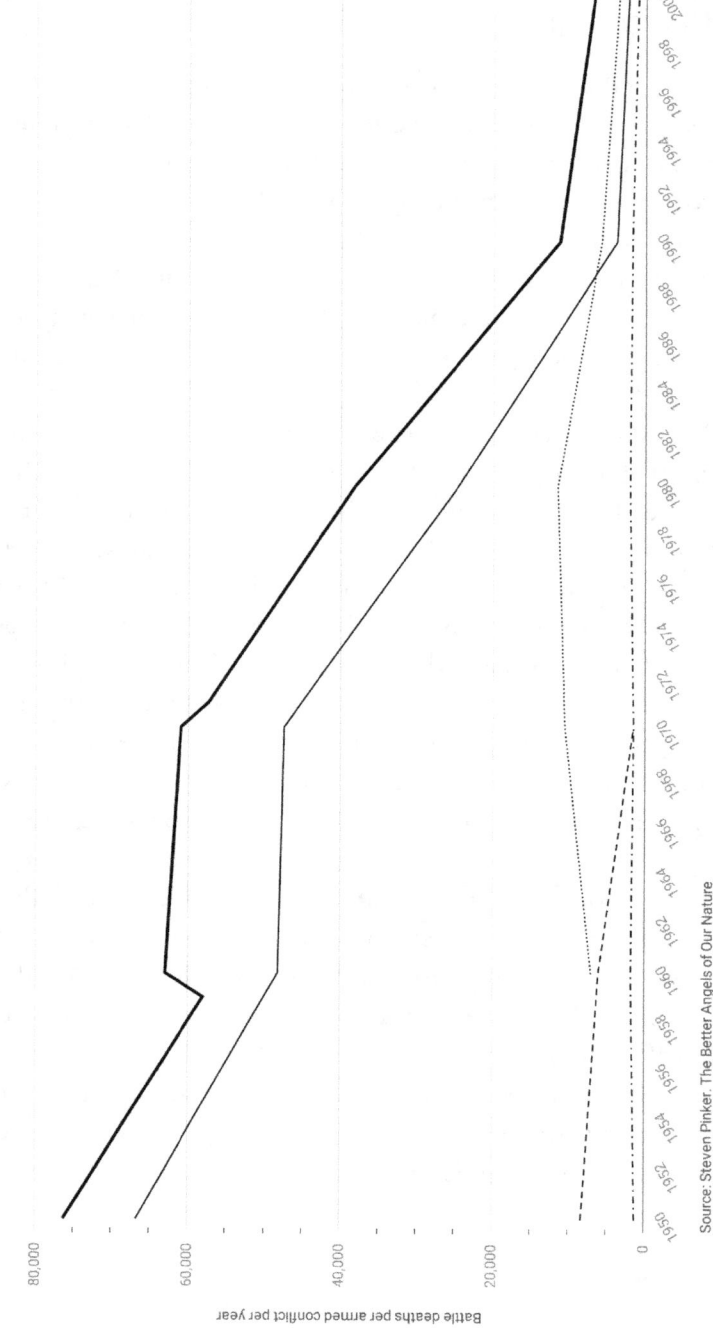

Deaths from Wars and Civil Wars

Battle deaths per armed conflict per year

80,000

60,000

40,000

20,000

0

1950 1952 1954 1956 1958 1960 1962 1964 1966 1968 1970 1972 1974 1976 1978 1980 1982 1984 1986 1988 1990 1992 1994 1995 1998 2000

Source: Steven Pinker. The Better Angels of Our Nature

Interstate wars · Internationalized civil — Colonial wars · Civil wars — TOTAL

Deaths Due to Ongoing Violent Conflicts

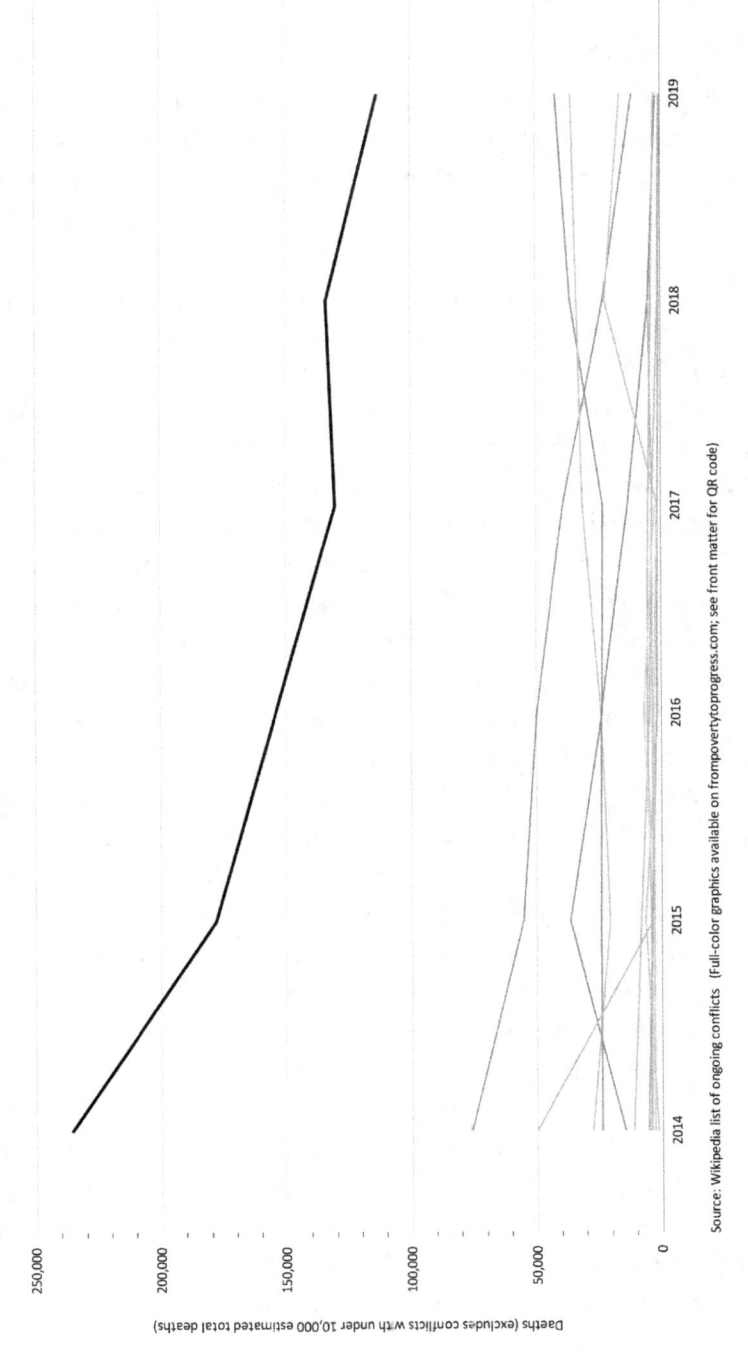

Deaths (excludes conflicts with under 10,000 estimated total deaths)

| 250,000 | 200,000 | 150,000 | 100,000 | 50,000 | 0 |

2014   2015   2016   2017   2018   2019

Source: Wikipedia list of ongoing conflicts   (Full-color graphics available on frompovertytoprogress.com; see front matter for QR code)

— Syria — Afghanistan — Mexico — Iraq — South Sudan — Yemen — Nigeria — Somalia — Sudan — Pakistan — Libya — TOTAL

Happiness

So far we have traced a number of metrics that show a clear trend of progress over the past few generations. The question remains, however; does all this material improvement affect people's happiness? Does better diet, health, education, sanitation, and standard of living, and lower levels of violence lead to greater happiness?

The World Value Survey enables us to measure levels of happiness in a concrete way. The poll started in 1981 and currently covers just under 100 countries. The following charts are based upon responses to the following question: "Taking all things together, would you say you are Very happy, Rather happy, Not very happy or Not at all happy". The figures display the total percentage of respondents who reported themselves to be "Very happy" or "Rather happy".

Some people, including me, are somewhat skeptical about how accurately these answers track with "real" happiness. While self-reported happiness is not a perfect measurement, it is closely correlated with other factors, such as frequency of smiling, rating of one's happiness by friends, and activity in the parts of the brain associated with pleasure. Whatever its limitations, self-reported happiness is measuring something real.

Among the Wealthy 12, 90% or more people in every nation but Germany, answered that they were either Very Happy or Rather Happy. With the exception of Germany, which has a clear upward trend and the United States and Australia, which have somewhat of a downward trend, there is relatively little variation over time. So while not all respondents agree, the high standard of living of the Wealthy 12 has translated into widespread happiness.

The slight downward trend since the 2008 recession is disturbing, though. The magnitude is small (a decline on average from 92% to 91%), but it is clearly in the wrong direction. It is possible that wealthy nations have hit "Peak Happiness", or it could be just a temporary dip due to economic recession. Either way, this short-term trend is trivial in comparison to the very high rates of self-reported happiness within the Wealthy 12.

In 1994, when the first data come from, the nations within the

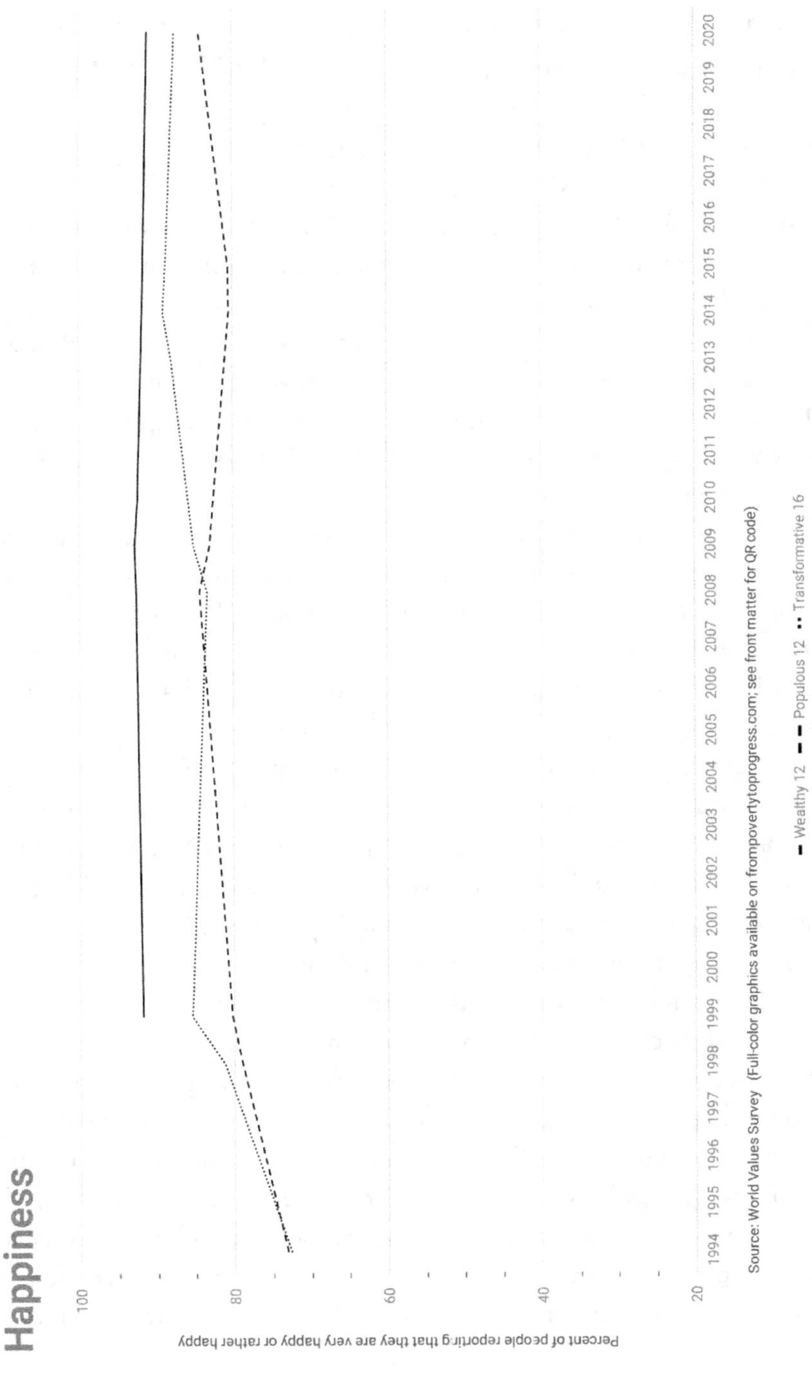

Happiness

Percent of people reporting that they are very happy or rather happy

100

80

60

40

20

1994 1995 1996 1997 1998 1999 2000 2001 2002 2003 2004 2005 2006 2007 2008 2009 2010 2011 2012 2013 2014 2015 2016 2017 2018 2019 2020

Source: World Values Survey (Full-color graphics available on frompovertytoprogress.com; see front matter for QR code)

Wealthy 12 — ■ ■ Populous 12 ∶∶ Transformative 16

Populous 12 had levels of happiness significantly below the Wealthy 12. They ranged
from a high of 81% in Turkey to a low of
65% in Nigeria. From 1994 to 2020 average
levels of self-reported happiness increased

KEY INSIGHTS

Progress has translated
into greater levels of
self-reported happiness.

from 73% to 84%. Only Egypt and Iran experienced a downturn. Bear
in mind that both of these nations experienced significant amounts of
political turmoil during that period.

With the obvious exception of Egypt, levels of self-reported happiness in all these nations increased to above 80%. Progress in the
Populous 12 has translated into increased self-reported happiness. Even
Egypt has experienced an upturn since 2014 that has almost completely
erased the previous downturn.

Unfortunately, the World Values Survey does not collect data on
many of the countries within the Bottom 20 (so I am not including the
graphic). Only Rwanda has multiple data points. Rwanda does show
an increase in self-reported happiness, from 86% in 2009 to 90% in
2014. With so little data, however, it is impossible to draw any conclusions on how happiness levels changed in these countries.

The Transformative 16 saw a strong increase in self-reported happiness from 1994 to 2020. In 1993, the happiest nation was Japan at
78% and the least happy was China at 67%. By 2020, all nations but
India had happiness levels of 80% or better. Only Puerto Rico failed to
show an upward trend. Recent drops in happiness in Hong Kong and
Thailand are likely related to political instability.

It is important to note that some of the happiest nations —
Singapore and Spain — do not show an upward trend, but polling
years are well after these nation's periods of rapid growth, so these
results do not invalidate the hypothesis that strong economic gains lead
to increased happiness. All other nations in the Transformative 16 with
more than a decade of polling data show increased happiness.

While we are missing data from the poorest nations, it is clear
that all of the positive trends documented previously do, in fact, result
in increased self-reported happiness. All of the Wealthy 12 nations

have very high levels of self-reported happiness and virtually all of the Populous 12 and Transformative 12 show substantial increases in self-reported happiness. Many of those nations are getting close to the levels of happiness in the Wealthy 12. While political instability and economic recessions clearly have short-term effects on self-reported happiness, the long-term trends are positive.

The field of happiness research is rapidly growing and somewhat beyond the scope of this book, but I would like to summarize some key findings from the literature:

- People from richer nations report being happier than people from poorer nations.
- Within nations, people with higher incomes report being happier than people with lower incomes. This holds for all 84 nations where we have data (with Mali being a possible exception).
- As the average income of a nation grows, so does the average level of happiness of its people.
- Economic growth leads to declining levels of happiness inequality, even if income inequality is also increasing. In other words, economic growth benefits the unhappy more than the happy. This is probably due to public goods that benefit everyone regardless of their income: better water, sanitation, education, health care, lower violence, etc.
- People tend to underestimate the level of happiness of other people within their society.

Conclusion

So what are we to make of all this data? I think a few conclusions are in order here:

Striking progress has occurred over the last few generations. While there are differences in the timing and magnitude, every metric shows clear progress. Each of the four categories of nations — the Wealthy 12, Populous 12, Bottom 20 and Transformative 16, show clear progress.

That progress has touched virtually every nation on the planet. While there are certainly individual nations that fail to show progress on individual metrics, the overall trend is clear for almost every nation. Even within the Bottom 20, evidence of progress is striking over the last three decades. And when we examine individual nations using broad metrics such as the HDI, it is clear that almost all nations share in that progress.

That progress has been evident in many different domains. The trend of progress is visible in every one of our metrics: economic growth, freedom, poverty, agricultural production, diet, sanitation, drinking water, life expectancy, neonatal mortality, education, housing and happiness.

Undoubtedly other domains have experienced progress worldwide — for instance, communication, transportation and entertainment come immediately to mind — but these are not documented in this study because we are missing solid data.

The broad-based improvement in so many metrics cannot possibly be explained away by increased inequality. Except for Per Capita GDP, virtually all of these metrics cannot be explained away with "yes, but the benefits all went to the rich." Most of these metrics were constructed in such a way that one poor person is equally weighted with one rich person. For example, it is inconceivable that the Percentage Access to Improved Sanitation goes up significantly while all the toilets are being monopolized by the rich. The metric only improves if a higher percentage of the population receives the benefits. The fact that rich houses may have more toilets than poor houses does not invalidate the enormous progress of poor people having access to improved sanitation. The same goes for Percentage Access to Improved Drinking Water and Percentage Access to Electricity.

Many of these metrics, such as democracy, freedom, famines, frequency of wars, and deaths from wars roughly equally benefit all members of society. One can make the case that they help the poor more as the associated problems hurt the poor more. For others, it is difficult to see how the rich could have monopolized all the gains.

How, for example, could the rich have eaten all the increased calories, benefitted from all the increased longevity, experienced all the increased happiness or accounted for all the increased years of schooling and literacy?

KEY INSIGHTS

There is no indication that progress is slowing down.

For almost all of these metrics, an improvement in the standard of living of the rich alone would have no positive outcome on the metric. It is only when a large segment of society experiences progress that the metric will change. So while there is no doubt that the rich benefit from progress, it is equally true that the rest of society benefits far more. Progress can be thought of as spreading a lifestyle that only the rich could previously afford to the rest of society.

Progress has translated into greater levels of self-reported happiness. Some critics of progress argue that an increased material standard of living is irrelevant because it does not translate into happiness. We have seen clear evidence that it does. Wealthier nations and those that do well on the metrics presented in this chapter have higher levels of self-reported happiness than poorer nations and those nations that do poorly on the metrics. As nations increase their wealth and improve on these metrics, self-reported happiness goes up. Even within nations, other studies show that people with higher levels of income report higher levels of happiness.

Everyday life for the vast majority of human history was tragic: people died young, lost a large percentage of their children, drank water that was polluted by feces, could not read, write or do arithmetic, had no schooling, received barely enough food to survive, and were run by tyrannical governments. Should we be surprised that ending or at least diminishing these terrible circumstances leads to increased happiness? No, of course not.

There is no indication that progress is slowing down. While there is the possibility that the wealthiest nations may reach a natural ceiling on progress, there is no indication that they are nearing it. Even in metrics where the Wealthy 12 measures 100% (or 0%), there is still reason to believe that improvements are continuing.

Most of the metrics that I use in this chapter were selected because they are typically used to measure progress in developing countries. In other words, they are specifically selected to identify things that wealthy nations typically have and poor nations typically do not have. This leaves open the possibility that there is a hard cap on progress and the Wealthy 12 have reached it.

While I do not go deeply into this topic in this book, progress can continue even after reaching the current levels of the Wealthy 12. For example, in the Wealthy 12, the rate of university attendance is increasing, the quality of sanitation and drinking water is improving, reading skills are improving and residents have greater access to varied foods. These improvements do not show up in the metrics above, but they are signs of progress nonetheless.

More importantly, middle- and low-income nations are mostly on a clear path to progress. It is quite conceivable that nearly every nation in the world will have development metrics matching the current levels of the Wealthy 12 within this century. On many metrics, a generation may be sufficient.

The progress metrics are closely correlated. One thing worth noticing is that all of these metrics are closely correlated. While I do not perform sophisticated statistical analysis in this book, it is clear from visual inspection that nations that perform well on one metric are highly likely to perform well on almost all other metrics. And nations that perform poorly on one metric are highly likely to perform poorly on almost all other metrics.

A nation with a high percentage of flush toilets is very likely to also be a nation with a high percentage of middle school graduates. A nation with high levels of neonatal mortality is also very likely to do poorly on adult literacy.

In addition, nations that show improvement in one metric are also highly likely to show improvement on almost all other metrics. There are clear variations between nations over time, but in general, all these metrics tend to move together. This is good evidence for the concept of progress.

Change in per capita GDP roughly measures progress. Also very important to notice is that all these metrics are closely related to per capita GDP. Per capita GDP loosely measures the standard of living, while change in per capita GDP loosely measures progress. This strongly suggests that economic performance is closely linked to progress.

> **KEY INSIGHTS**
>
> Dramatic levels of progress can take place within one generation.

By creating wealth, societies can purchase the technologies, learn the skills and establish the social organizations that lead to improvements in all of these metrics. The close correlation between all these metrics of progress to per capita GDP also makes it much easier to chart progress across the globe over the last 1,000 years when we do not have data on specific metrics.

Dramatic levels of progress can take place within one generation. While progress often takes place slowly over many generations, under the proper conditions extremely rapid progress can take place in just a generation. The Transformative 16 all experienced at least a doubling of per capita GDP within 20 years. Given that nations can benefit by merely copying what wealthier nations have done, it seems likely that many of the poorest nations today may surpass that level of improvement soon.

While poorer nations have shown striking progress, they seldom surpass rich nations. Poor nations have instead narrowed the gap between themselves and rich nations. The progress among the Bottom 20, Transformative 16, and Populous 12 has been stunning, but it is interesting to note that few of them have been able to surpass the Wealthy 12. Nations that ranked highly on metrics in 1950 still rank highly today. Most of the nations that ranked low in 1950 still rank low today. Some of the Transformative 16 nations have been able to jump up in their relative ranking, but this is less common than a steady improvement of all rankings together.

I guess that this is due to the difference between innovating and copying. The Wealthy 12 have innovated a large number of solutions to the problems identified in our metrics. Because of those innovations, their standards of living have reached unsurpassed levels. Since they have evolved into societies that are conducive to innovation, they can remain at the forefront of progress.

The rest of the world has largely improved by copying those innovations made by the Wealthy 12. A nation can rapidly catch up to the innovating nations by copying the Wealthy 12, but that is not a sufficient strategy for actually surpassing them. To do so, the rest of the world has to learn to innovate as effectively as the Wealthy 12. This, as we will see, is a much bigger challenge.

This also makes it easy to understand why some people do not believe in progress. Critics of progress typically focus on inequality, which still exists and probably always will. But it looks different if one shifts the focus towards the overall trend of the stunning progress that has taken place, and the fact that this progress has benefitted the poor more than anyone. The fact that the wealthy have higher levels of luxury consumption does not invalidate the far more important trends that mean that the poor have higher levels of the basic goods and services necessary for survival.

A world where virtually every nation has levels of standard of living similar to the Wealthy 12 today is achievable. China in 1950 had a per capita GDP of $637. Currently 19 out of 20 of the Bottom 20 nations have higher levels than that. In 2016 China had a per capita GDP of $12,569. That is higher than every one of the Wealthy 12 nations in 1940, except Switzerland. And China has grown significantly since 2016.

This level of progress would have been inconceivable to Westerners in previous decades, but it has nonetheless happened. Japan followed the path blazed by the West, and then China followed Japan. More countries are currently following China. And each country that follows this path to progress makes it easier for other countries to do the same, as they have more examples to copy.

One can even extrapolate these trajectories onto Sub-Saharan Africa, the poorest region in the world. It is quite possible that this region could become almost as affluent as the Wealthy 12 are today within this century. Suburbs full of nice houses with two-car garages for most people living in Congo is quite achievable. It is not guaranteed, of course, but achievable. No new technologies, skills or social organizations need to be invented. Sub-Saharan Africa can theoretically just copy what worked for dozens of other societies.

KEY INSIGHTS

A world where virtually every nation has levels of standard of living similar to the Wealthy 12 today is achievable.

So, progress is real. And this progress is even more stunning when we put it into context with how people lived in previous centuries. So what caused all of this progress to occur?

THE FIVE KEYS
TO PROGRESS

In the previous chapter, I made the case that we live in a world of progress, and that progress plays an important role in making all of our material lives better. In the next two chapters, I explain what caused this progress to take place and how it works today.

First, I will quickly outline the concept of Five Keys to Progress, and then explain why they are important. Then I will go into more detail on each of the five keys.

I believe that the Five Keys to Progress is an essential unifying concept for understanding progress. They are critical because they are the **necessary preconditions** for a society changing from a state of poverty to a state of progress, and they are **actionable** in today's world. In other words, the concept not only helps to understand the world but also how to make it better.

The Five Keys to Progress enable us to cut through all the clutter of history and modern times so that we can focus on what really matters. They enable us to answer some of history's most difficult questions, as well as providing policy solutions and practices that can work.

Let's Dismiss Some Common Fallacies

First, let's dismiss some commonly held views about the causes of our current progress.

This progress did not come from the government. While governments do play some role in progress, particularly in education, infrastructure and basic health care, it is not a dominant one. And as we will see later, throughout history governments have done far more to hinder progress than they have done to promote it. Governments may redistribute the benefits of progress, but they do not create it.

Nor has political activism played a major role in progress. Certainly, political movements to overthrow authoritarian regimes that hinder progress have played an important role. But they do so by removing obstacles to progress, not by actually creating progress. And once basic liberties and democratic governance are established, the benefits of political activism with respect to promoting progress rapidly diminish.

Nor can we see any political leaders who have played important roles in this progress. Certainly, some political leaders have made their nations a better place, but their impact overseas is usually quite limited. Leaders primarily promote progress by enabling other people to create progress.

We cannot point to any pieces of legislation or the establishment of new programs that played an important role in promoting progress. Certainly, some legislation or programs in individual nations have helped, but the metrics are too widely dispersed across the world and legislation has been too diverse between nations to account for much of the change.

Nor has redistribution, in general, played an important role in progress. Wealth must be created before the benefits can be redistributed. If progress does not exist, then there is not much to redistribute. The people participating in progress have done far more to benefit individuals than any redistribution that they might receive.

All of the political factors mentioned above can play a role in breaking down barriers to progress, but they must rely on other factors

> **KEY INSIGHTS**
>
> The Five Keys to Progress are the necessary preconditions for a society to change from a state of poverty to a state of progress.
>
> Progress does not come from government, politics or policies.

to deliver progress. Rather than flowing down from politics and government, progress bubbles up from society.

Progress Is An Evolutionary Process

Progress is an evolutionary process. Evolutionary processes are not controlled or directed by any government, organization, person or groups or God.

Progress is the outcome of millions of small-scale decisions made by individuals who were trying to solve short-term local problems. Few of them were trying to make the world a better place. Few were even thinking about the world as a whole. Few of these people were involved in politics or government.

I will go into much greater detail in later chapters, but I want to expand your view of the "causes" of progress. Because progress is an evolutionary process, its causation is tricky to define. It is a bit like debating the cause of life on Earth (another evolutionary process). There are many "causes" that interact with others in very complex ways.

I believe that this complexity is part of the reason why previous researchers have not discovered the Five Keys to Progress earlier.

When viewed from a very high level, evolutionary processes require two things:

1. Necessary *pre-conditions* that enable lower-level factors to come into existence and survive and change over time. There are often many different pre-conditions that must come together to create the "Goldilocks conditions." At any time, the pre-conditions can disappear and the entire evolutionary process collapses.
2. Specific mechanics that explain *how* that evolution takes place. This usually consists of multiple mechanics that interact with each other in complex ways.

This chapter is about the necessary pre-conditions for a society to transition from a state of poverty to a state of progress. The next chapter is about the mechanics of how progress actually works.

Introducing the Five Keys to Progress

So let's reiterate the Five Keys to Progress. To transition from poverty to progress, a society needs to acquire the following preconditions:

1. **A highly efficient food production and distribution system.** This enables societies to overcome geographical constraints to food production so that large numbers of people can focus on solving problems other than getting enough food to eat.

2. **Trade-based cities packed with a large number of free citizens possessing a wide variety of skills.** These people innovate new technologies, skills and social organizations and copy the innovations made by others.

3. **Decentralized political, economic, religious and ideological power.** It is of particular importance that elites are forced into transparent, non-violent competition that undermines their ability to forcibly extract wealth from the masses. This also allows citizens to freely choose among institutions based upon how much they have to offer to each individual and society in general.

4. **At least one high-value-added industry that exports to the rest of the world.** This injects wealth into the city or region, accelerates economic growth and creates markets for smaller local industries and services.

5. **Widespread use of fossil fuels.** The incredible energy density of fossil fuels injects vast amounts of useful energy into society

enabling it to solve a wide variety of problems. Without this energy, life would return to the daily struggle for survival that dominated most of human history.

Each of the Five Keys to Progress is **necessary** for a society to transition from a state of poverty to a state of progress, but **none are sufficient by themselves**.

Origins of Progress

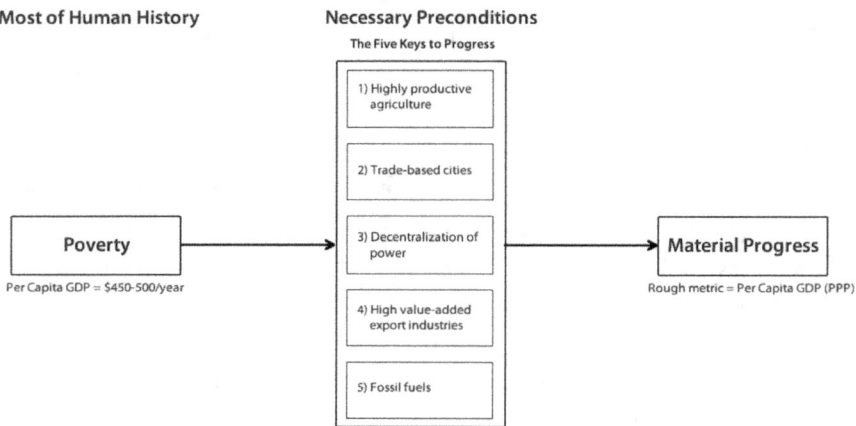

Once a society acquires the Five Keys to Progress, that society can transform itself into a vast, decentralized problem-solving network. Instead of people competing against each other for scarce resources such as food, status and land, individuals can focus on solving each other's problems at scale by cooperation.

Individuals quickly find that it is better to focus on solving each other's problems at scale in the marketplace rather than exclusively focusing on their own individual problems. Individuals can magnify their problem-solving abilities by forming new organizations that enable people to cooperate on a larger scale. They have the incentive and the desire to innovate technologies, skills and organizations as well as copying those that show positive results.

I believe that the degree to which peoples have enjoyed progress in history (and to a certain extent today) is largely determined by long-term historical factors that go back centuries or even millennia. These factors determined the extent to which societies acquired the Five Keys to Progress. For most of human history, there was no progress, because these five keys were either completely missing or were very underdeveloped.

In the distant past, a few Commercial societies in Northwest Europe acquired the first four keys, creating sufficient conditions for progress to occur *before* the Industrial Revolution. When one of those societies, Britain, added the fifth key, it created sufficient conditions for the widespread progress of the modern era.

As more and more societies copied those five keys, progress spread to new societies over time. First, it spread to other parts of Northwest Europe, then to North America, then to Germany and parts of Southwest Europe, and then to Japan. For a while after that, progress seemed to stop spreading.

KEY INSIGHTS

Each of the Five Keys to Progress is necessary for a society to transition from a state of poverty to a state of progress, but none are sufficient by themselves.

Once a society acquires the Five Keys to Progress, that society can transform itself into a vast, decentralized problem-solving network.

The degree to which peoples have enjoyed progress in history (and to a certain extent today) is largely determined by long-term historical factors that go back centuries or even millennia.

More recently, progress has spread to most of Eastern Europe, East Asia, South Asia and Southeast Asia. Over the last 20 years, significant progress has even been achieved in Sub-Saharan Africa.

Given the positive impact on so many different continents and cultures, there is every reason to believe that progress can spread through the entire world. Maintaining as much of those five keys as possible in as many different nations as possible is critical for preserving progress in wealthy nations and increasing progress in developing nations.

The Five Keys to Progress enabled humanity to move from "slow change with no direction" to progress that benefitted mankind. Once a society acquires the five keys, it can escape the poverty trap imposed by geography, demographics and elite domination. The masses can begin to enjoy a long-term increase in their standard of living. Indeed, human history itself can be viewed as a vast evolutionary process that led to the accidental discovery of the Five Keys to Progress.

As long as the five keys remain in effect, a society can solve an extraordinary number of problems. The result, in the long run, is widely-shared prosperity.

Key #1: Productive Agriculture

Cities, as we will see, are one of the Five Keys to Progress. But cities would not have been possible without a highly efficient food production and distribution system. For all practical purposes, this means agriculture. And not just any agriculture; it has to be highly productive agriculture.

Food has been the critical constraint on innovation and progress throughout the vast bulk of human history. I know that this fact is very hard for modern readers to relate to, but it is true. Today when we are hungry, we go to the refrigerator or pantry. When the refrigerator or pantry is empty, we go to the grocery store. Easy, right?

Well, no. Not for our ancestors.

While getting enough food to survive is easy in modern societies, it was an epic task for our ancestors. For the overwhelming majority of our ancestors, the quest to acquire enough food to survive took up the majority of their waking hours. It was an obsession, and all of society was organized around the most effective means to do so within the local environment.

Before one can innovate, one must first survive. In order to survive, one must eat large amounts of food. If one has to spend the vast majority of one's time focused on acquiring food, one cannot devote much time to innovating non-agricultural skills, technologies and organizations. Our ancestors effectively traded time for food.

As we will see in the later chapters, the vast majority of mankind devoted the bulk of their waking hours to the quest to produce enough food to eat. Very gradually over time, they innovated new technologies, skills and processes that enabled more effective means of producing and distributing food. This quest for food production

KEY INSIGHTS

Key #1: A highly efficient food production and distribution system.

was so all-encompassing that we can categorize entire societies by how they did so.

This quest for food greatly affected where we lived. Until the last few centuries, humans had to disperse geographically in order to acquire food, because the subsistence technology of the day was not productive enough for one family to grow enough of a food surplus to support urban populations. The endless drudgery of acquiring food has stifled the human potential for innovation and progress for millennia.

There is only so much food that can be acquired from any one acre of land with simple technology. With technological innovation, we have been able to radically increase this amount, but for any given natural environment and suite of technologies, there is a fixed amount of food that can be produced per acre.

So humanity dispersed to survive. They had no choice. Survival comes first.

But this dispersal undermined the ability of large numbers of people to interact regularly. Dispersal made it harder to copy technologies, skills and organizations. With far fewer models to copy, innovation and diffusion were far slower than they could have been.

To make things worse, the type of food that can be produced in any specific area is highly constrained. In some regions, fishing or hunting marine mammals is possible, but not in most areas. In other regions, hunting big game is possible, but not in most areas. In some areas, cultivating rice is possible, but not wheat or corn. In other areas, cultivating wheat is possible, but not other staple crops.

Since food production and distribution is not something that one individual can do, the entire society had to be sculpted around the type of food that could be produced. Technologies, skills, organizations and values of societies were all greatly affected by what type of food could be produced in their geographical environment.

What is more, each one of these food types had very different amounts of energy and nutrients relative to the human work effort. This meant that some societies had huge advantages over others, simply because they lived in regions with the most cost-effective food sources.

Some geographical regions simply could not support much food production, so they doomed their residents to be trapped in poverty. A very lucky few regions were blessed with geography that enabled far more food production per unit of work. These unique geographical characteristics left open the possibility of progress evolving within their borders.

These geographical differences still account for a significant proportion of the inequality between societies that exist to this day.

For thousands of years after the invention of agriculture, our food production systems were deeply unproductive by modern standards. Farmers struggled to support their own families. They needed to store seed for the next harvest, pay taxes to the state and rent to their landlord. They also had to leave half of their land fallow each year.

Traditional farmers lived in a highly uncertain environment where one bad harvest could lead to deaths in the family. They were also under threat from droughts, floods, marauding armies, epidemics, pests and famine. They could do everything right, and then one bad thing happened, and their family was in great peril.

Living in such a situation makes innovation very difficult. The downside of one bad harvest far outweighs the potential benefits of a slightly better harvest. Given their circumstances, farmers were very reluctant to experiment unless they have to do so. As David Grigg has pointed out in *The Agricultural Systems of the World*, virtually all agricultural systems have been relatively unchanged for millennia.

Fortunately, there was one important exception to this rule: Northwest Europe and the regions settled by Northwest Europeans. While other agricultural systems hardly changed for thousands of years, these regions have undergone many transformations. Each of these transformations resulted in a significantly increased food surplus per family of farmers.

Mazoyer and Roudart sketch out these transformations in exhaustive detail in their book *A History of World Agriculture.* I hope to go into much more detail on this topic in future books in this series, but I can do a quick recap here.

Ancient farmers of Europe and the Mediterranean used a relatively unproductive two-year rotation system along with scratch plows pulled by oxen. Medieval farmers gradually evolved a much more productive three-field system that made use of heavy plows pulled by horses. This likely doubled farm productivity, to a large extent because it reduced fallowed land from 50% to 33% (fallowed land is land that is not planted so the soil can regenerate nutrients for replanting at a later date).

The critical breakthrough occurred when farmers in Flanders (which is now part of Belgium), the Netherlands and probably Northern Italy figured out how to plant different crops so fallow land was no longer needed. This critical discovery created the food surplus that Commercial cities needed to survive and prosper.

Later, English farmers copied these innovations and invented tools that made the system even more productive. In later centuries American farmers learned how to mechanize and motorize cultivation. This pushed agricultural productivity into the stratosphere.

This increased productivity created a food surplus that could be profitably sold to cities. Those that were lucky enough to live near rapidly growing cities had a strong incentive to specialize and export their

KEY INSIGHTS

Food has been the critical constraint on innovation and progress throughout the vast bulk of human history.

For the overwhelming majority of our ancestors, the quest to acquire enough food to survive took up the majority of their waking hours.

Before one can innovate, one must first survive. In order to survive, one must eat large amounts of food.

Humans dispersed to survive, but this dispersal undermined their ability to innovate and copy the innovations of others.

produce to the cities. In the Medieval system, relatively few crops were exported. Wine and wool were the major exceptions. Now, in areas near commercial cities, a dense web of economic connections knitted together cities and the surrounding countryside for many miles around.

Products that were previously only consumed on the farm could now be exported to the cities. An entire secondary market for agricultural goods came into being. Some of these goods were raw foods for human consumption — beef, pork, mutton, lamb, chicken, milk, eggs, nuts, honey, fruits and vegetables.

Other agricultural goods were processed for human consumption — oil, cheese, butter, cream, yogurt, beer, wine, whiskey, gin and rum. Some were goods processed from plants — flowers, woad, hemp, rope, wool, flax and linen. Others were goods processed from animals — leather, manure, bone, sinew, marrow, feathers and fur.

Some of the goods were processed on the farms, while others were transported to the cities for further processing and sale. Previously, this type of commercialized agricultural exports was limited to wine and wool. Now it had spread to a suite of other agricultural products.

As farmers increased their exports to neighboring cities, some found it more profitable to specialize in one agricultural good. Because it was now far easier to buy food in the market, it was no longer necessary to grow every necessary foodstuff on your own farm. Farmers were now free to focus their time on one product, giving them an incentive to increase their skills, adopt new technologies and techniques and expand the scale of their production.

This increased specialization of crops enabled the Northern Italian, Flemish, Dutch and English farmers to do what would previously have been unthinkable: to shift away from growing grain. Because grain

> **KEY INSIGHTS**
>
> Food production is highly constrained by geography.
>
> The entire society had to be sculpted around the type of food that could be produced in the region.
>
> Geographical differences still account for a significant proportion of the inequality between societies that exist to this day.

was one of the few crops that could be stored
and transported over long distances, Flanders
found it more profitable to import large
amounts of grain from the Baltic.

This grain trade proved highly profitable
for Flemish and Dutch merchants. And having the option to buy cheap imported grain
gave Flemish and Dutch farmers even greater
incentives to shift their production to more
profitable secondary agricultural goods.

Secondary agricultural products created
enormous benefits to cities as well. With a
flood of products that started out as domesticated animals and culti-
vated plants and were then processed into profitable goods, a radical
diversification of urban economies could take place. Entire new professions, skills and industries sprouted up.

This not only created a wide range of producers, but it also created a
large growth in the number of consumers. Each specialized worker in the
city had no choice but to buy what they needed on the market. Whereas,
they had previously been forced to produce all their own goods, now they
were modern consumers who purchased those items on the market.

This created a positive feedback loop that stimulated the expansion of markets. The more city dwellers found it necessary to purchase
goods on the market, the more farmers could specialize in the most
profitable products. The more farmers focused their skills and energy
on these products, the more the price dropped. This stimulated more
urban demand, which continued the feedback loop.

Proximity to growing cities also made it more realistic to seek seasonal employment in the cities during times of low labor needs. Since
farmers were no longer hovering on the edge of survival, they could
afford to seek out the most highly paid non-agricultural jobs. Some of
these temporary migrants decided to stay in cities permanently. This
constant flow of migrants kept cities growing in population.

The greater variety of foods also enabled both farmers and urbanites

KEY INSIGHTS

While other agricul-
tural systems hardly
changed for thousands
of years, Northwest
Europe has undergone
many transformations.
Each of these transfor-
mations resulted in a
significantly increased
food surplus per family
of farmers.

to eat healthier and more balanced diets. The Medieval diet was largely restricted to bread and beer. Now meat, eggs, dairy products, fruits and vegetables could all be consumed in much greater quantities. While it was hardly a diet that modern nutritionists would recommend, it was far superior to the diet of Ancient or Medieval farmers.

While the new no-fallow system created enormous advantages for farmers and urbanites, the system was very slow to spread. Only regions with a large number of cities and widespread peasant ownership of land could realize its benefits. Resistance to change from peasants, nobles and kings was substantial until the 19th Century.

Only Flanders, Netherlands, Southeast England, Southern Scandinavia and scattered parts of western and northern Germany and northern France saw a widespread transition to the new agricultural system. As late as 1840, the vast majority of European farmers were still using the same farming system that they had used in 1300. Most of the adoption of the no-fallow system coincided with the Industrial Revolution in the middle and late 19th Century.

So agriculture was not some sideshow to progress. It was and still is one of the key foundations that progress is built upon. Today, we get caught up in flashy innovations on the leading edge of innovation, and we should. Those innovations are very important. But always remember that all of these innovations rest on the toils of the lowly farmer.

Shifting back to modern times, none of this means that a growing economy cannot import food from overseas. Many, in fact, do so. It does mean that virtually all societies that have transitioned from poverty to progress did so by increasing the productivity of their own agriculture beforehand.

Doing so does not create progress in itself. It merely makes progress possible. The actual progress occurs in the cities, but city-dwellers are dependent upon food imports from the countryside.

City dwellers still need to eat and, the more that people migrate to the cities, the fewer farmers there are to produce a larger food surplus. City dwellers in poor regions do not have the money to import food because it is too expensive. They need locally-produced food, and, more importantly, they need cheap local-ly-produced food. This is why a region cannot get rich without big increases in productivity in the agricultural sector.

A handful of societies have been able to survive and prosper from food imports before they industrialized. Hong Kong and Singapore today are two clear examples. Both are city/states that had no other choice due to a lack of land.

Some prosperous nations of the past sup-plemented their agriculture with fishing. The Dutch Republic and pre-Industrial England are two examples. This is a bit like importing food, but I still consider this to be domestic food production. Very few nations in history imported substantial amounts of food *before* they experienced progress.

> **KEY INSIGHTS**
>
> Productive agriculture made trade-based cities possible.
>
> Trade-based cities gave local farmers incentive to spe-cialize and increase production.
>
> Early industry was based upon products from farms.

And notice the wording on the first key. I deliberately added "and distribution system" to cover the exceptions. To import food, one needs an excellent distribution system, usually in the form of shipping. Even societies that rely on domestic food production must have a good food distribution system to store and transport the product to cities. Most typically this will be in the forms of carts, wagons or barges on canals.

It is absolutely true that many societies that transition from poverty to progress import sizable amounts of food *after* they make the transi-tion. With the increased wealth, their citizens naturally want a greater variety of food to enjoy, and they have an incentive to free up labor from farms. But this is rarely the case before the transition.

That is why productive agriculture is the first, and most important, key to progress.

Key #2: Cities

In *Triumph of the City*, Edmund Glaeser calls cities "our species' great-est invention." While this might be a bit of an exaggeration, Glaeser is onto something important. Most people rarely stop to consider the importance of cities to human history.

City dwellers are invariably wealthier than residents in rural areas, often dramatically so. This is because of the city's unique ability to foster innovation and diffusion. Humans as individuals are undoubtedly more intelligent than animals. But the differences in intelligence between humans and chimps or dolphins are not enough to account for how far humans have diverged from other animals in their evolution.

It is only when large numbers of humans live and work in close geographical proximity that humanity's greatest advantage emerges. Increased density of interaction between humans creates a network of skills and social organizations that is far beyond what any other species has been able to achieve.

Humans in large groups have the ability to cooperate, specialize, learn new skills, copy ideas, test those ideas and constantly improve them based on feedback from other humans. While humans in small groups tend to use the same technology for generations, humans in large groups can become a network that churns out new ideas and tests them for success in a rapid manner. And when those large groups of people are in constant contact with other large groups of people, progress is even more rapid.

The large, dense populations within cities also naturally promote specialization, which leads to a broader skills base. A small town might have one restaurant. Because it is the only restaurant in the area, it must serve a cuisine that is acceptable to most of the town.

As the population size of the city increases, the number of restaurants naturally increases as well. But something even more important takes place. When there are a dozen restaurants in a town, each one of them can cater to the tastes of a different segment of the population. One can serve Chinese cuisine, one can serve pizza and another can specialize in breakfast. Increase the population size to millions and suddenly you can have thousands of restaurants, specializing in increasingly narrow markets. An increase in population promotes increased specialization. And each specialization requires a different set of skills.

As we will see later, technological innovation is exponential, because it is partly based upon the number of technologies already in existence. The larger the number of technologies, the larger the number of potential combinations for new technologies in the future.

But technologies do not combine themselves. Humans have to do that. And while isolated individuals can try a certain number of combinations, a large number of people living in close proximity are far more likely to try all possible combinations that their current technology base allows. And because technologies can only come into being when there are people with the necessary skills to make that happen, those people need to have a wide variety of skills.

Cities matter because they concentrate large numbers of people into a small area. Whereas rural areas of the past had low-density populations with people focused on skills related to growing food, cities had high-density populations with people specializing in a wide variety of occupations, each with unique skills, technologies and social organizations.

And these urbanites were in constant daily contact with each other, giving them the ability to copy from a wide variety of people. Cities enabled people to view how other people worked, view their technology and perhaps share ideas. Most of those people who they shared with would be in their family or occupation, but some might be strangers engaged in other occupations.

Cities have always been havens of freedom. This is important because free people enjoy the benefit of their skills and innovations, while slaves have no incentive to innovate. If a slave innovates something new, only the master benefits. While most pre-modern societies had some form of forced labor (slavery, serfdom, peonage, etc.), trade-based cities make use of free labor in the marketplace. Forced labor could simply not compete with free labor in occupations that require skill, innovation and learning.

But not all cities are created equal. Some cities innovate at a far higher rate than others. As more and more trade-based cities evolved, some of them became the locus of emerging technologies. Important cities in history include those in Northern Italy during the Middle Ages (Venice and Florence were the most important), Bruges and Antwerp in modern-day Belgium, Amsterdam from about 1580-1670, London, New York City and today's Silicon Valley.

These trade-based cities have played an enormous role in technological and organizational innovation because they had heavy concentrations of people with skills related to emerging technologies of their day. For almost 1,000 years, these cities have been the engines of progress.

Except for innovation in the field of agricultural technologies and the extraction of minerals, virtually all innovation has come from cities. This is quite extraordinary, given how few people lived in cities until recently. For a good 100,000 years, cities did not exist. Hunter-Gatherers congregated in seasonal camps or fishing villages when and where food sources were highly concentrated, but they had at most a few hundred inhabitants.

Even after cities evolved, a very small proportion of the population lived within them. In most Agrarian societies, cities with a population of over 10,000 inhabitants made up less than three percent of the population. And compared to modern standards, a "city" with a population of 10,000 would be considered more of a village today.

> **KEY INSIGHTS**
>
> Virtually all innovation happens in cities.
>
> Humans living in close proximity can cooperate, specialize, learn new skills, copy ideas, test those ideas and constantly improve them based on feedback from other humans.
>
> Until recently, less than 3% of people lived in cities.

Key #3: Decentralization of Power

So productive agriculture leads to cities, and cities lead to innovation and progress. Sounds pretty simple. Unfortunately, it is not so easy. Whenever farmers create a food surplus that can potentially lead to the growth of dynamic trade-based cities, elites in those societies have other ideas.

KEY INSIGHTS

Key #3: Decentralized political, economic, religious and ideological power. It is of particular importance that elites are forced into transparent, non-violent competition that undermines their ability to forcibly extract wealth from the masses.

Unfortunately, throughout most of human history, the bulk of the food surplus has been extracted by political, economic or religious elites in the form of taxes and land rents. Rather than allowing specialists in cities to consume this food surplus, the elites spent the food surplus on conspicuous consumption, military conquest, signaling their social status and celebrating their religious or ideological visions. These elites effectively stifled the growth of trade-based cities, which in turn stifled the possibility of progress.

Sometimes these elites extracted wealth from the peasantry individually, as in European feudalism, but more often elites established centralized extractive institutions to do so on a vast scale. Usually operating as government-sanctioned monopolies, these extractive institutions channeled the food surplus generated by farmers towards elites. Unfortunately, through most of human history, the more productive farmers have become, the more extractive institutions funneled that wealth to elites.

And even worse, elites funneled this food surplus into building powerful military machines that competed against each other to expand the scope of extraction into neighboring areas. The Chinese, Roman, French, Ottoman, Persian, Spanish and Portuguese empires are just a few of the dominant empires that have chosen this path. Many other potential empires attempted to do the same, but they were outcompeted by their more famous competitor empires.

For this reason, the decentralization of political, economic, religious and ideological power is essential to innovation and progress. Ideally, this decentralization comes from the creation of institutions that compete against each other without the use of violence. When institutions compete peacefully, they can no longer acquire all their resources by extracting from farmers and urbanites.

Organizations that are forced to compete non-violently have the incentive to offer material benefits to the masses in order to acquire more resources. The people are no longer beasts of burden to be exploited, but potential customers, employees, investors and voters.

Organizations that compete non-violently have a strong incentive to embrace new technologies, skills and processes that give them a competitive advantage against other institutions. At this point, instead of having the incentive to stifle progress, elites suddenly had the incentive to promote progress.

Today we often hear activists complain about the entrenched power of elites, but those activists miss the point. Since mankind evolved past Hunter-Gatherer societies, there have always been elites.

The key question is whether elites acquire their wealth from extracting wealth using government-sanctioned monopolies that rest upon violence, or whether elites are forced to compete against each other non-violently. When elites are forced to compete against each other non-violently, they must offer something to others to win that competition. This competition gives city-dwelling specialists a sphere where they can innovate new technologies, skills and organizations without being stifled by extractive institutions.

Modern societies have evolved a number of means to force elites to compete against each other non-violently. Political parties, rule of law and elections force political and ideological elites to do this. So

> **KEY INSIGHTS**
>
> Unfortunately, throughout most of human history, the bulk of the food surplus has been extracted by political, economic or religious elites in the form of taxes and land rents.
>
> When elites are forced to compete against each other non-violently, they must offer something to others to win that competition.

do markets, property rights and corporations. Meanwhile, separation of church and state and the concept of religious liberty forces religious elites to compete non-violently against each other for worshipers.

And the specialization of institutions in a modern society means that an institution can compete in only one of those spheres. They must specialize in one of politics, economic or religion. When elites compete non-violently, the rest of society has the opportunity to choose which sub-section of the elite most benefits society. And with so many options, they can mix and match as they choose.

> **KEY INSIGHTS**
>
> Rather than conquering new lands or squeezing taxes from the peasantry, modern elites become wealthy by creating wealth. They do so by innovating new technologies, skills and social organizations. The innovators gain vast wealth from those innovations, but the masses as a whole receive far more of the benefits.

Today, we take all of that competition for granted, but for millennia Agrarian societies (like authoritarian regimes today) strictly limited competition by imposing government-sponsored monopolies. These monopolies enabled political, economic and religious elites to extract wealth from the masses. More recently ideological elites have played that role. Any new organizations that could create wealth that will benefit the masses are a distinct threat to elite power, as they could become power bases for rivals.

Forcing elites to compete non-violently against each other is critical because it changes how people become wealthy. Rather than conquering new lands or squeezing taxes from the peasantry, modern elites become wealthy by creating wealth. They do so by innovating new technologies, skills and social organizations. The innovators gain vast wealth from those innovations, but the masses as a whole receive far more of the benefits.

Key #4: Export Industries

Regardless of how productive agriculture or innovative cities are, for progress to take place, a society must have at least one high-value-added industry that exports to the rest of the world. These industries inject wealth into the region and accelerate economic growth. This wealth can then be spent locally by its employees, generating demand for a gaggle of smaller local businesses. They also create a revenue stream for governments to invest in education, health, transportation, sanitation and energy infrastructure.

By exporting to the rest of the world, the industry radically increases the potential demand for their goods. If a farm or city is restricted to customers within their own borders, its economy has far less potential for growth. And the more value that the industry generates, the higher the potential profits. That is why high-value-added industries that are competitive enough to export to the rest of the world are so critical to promoting progress within a region.

The nature of the industry that is needed varies greatly over time. In the distant past, a mineral or crop might be sufficient. More typically today, it requires some form of manufacturing. Textiles, steel, and consumer electronics are all industries that played critical roles in various nations during their initial industrialization.

To successfully export, a city needs the necessary technology, skills, organizations and capital. These factors are typically acquired by copying them from richer regions that already have them, modifying them for the local environment and then slowly learning by doing. Often skilled immigrants from richer nations play a critical training role in the learning of new skills.

The emerging discipline of Economic Complexity gives us the best understanding of how this works. Cesar Hidalgo and Ricardo Hausmann have played pioneering roles in this field. The best introduction to their views is in the book *The Atlas of Economic Complexity: Mapping Paths to Prosperity.*

Hidalgo and Hausmann argue that modern societies acquire productive knowledge by distributing that knowledge among many specialized workers. Organizations and markets then combine that knowledge to make useful products. Skills needed for industries can only be taught face-to-face making knowledge transfer very difficult. If an industry is missing only one key skill, it cannot be competitive.

This places poorer nations in a "Catch-22" situation. They cannot create high-value-added industries until they acquire the necessary skills, but they cannot acquire the necessary skills until they already have a functioning industry. This creates a fundamental gap that developing nations have difficulty bridging. On the face of it, this appears to make economic growth impossible. Fortunately, we know from history that economic growth is possible.

KEY INSIGHTS

Key #4: At least one high-value-added industry that exports to the rest of the world. This injects wealth into the city or region, accelerates economic growth and creates markets for smaller local industries and services.

The theory of economic complexity gives us the key intellectual breakthrough that industries are related to each other because they share common skills. For example, manufacturing shoes is closely related to manufacturing hats because they share common production skills. Those same industries are far less related to manufacturing tractors or pharmaceuticals because those industries require very different skills.

And skills are not the only factor. Related industries also require similar technologies and organizational needs. This makes it theoretically possible for poor nations to leverage the limited knowledge that they have from current sectors of the economy to other related sectors that offer higher value.

Hidalgo and Hausmann use the analogy of a monkey traveling through a forest in search of food. The monkey is on one side of the forest with few bananas (limited export options) and wants to get to the other side of the forest with many bananas (a number of profitable high-value-added export industries). The monkey can only travel one branch at-a-time.

So what does the monkey do? He gradually moves to the nearest branch with more bananas than the one he is currently on. He then uses the energy gained from the bananas (i.e. the technologies, skills, organizations and capital) to go to the next branch. It is a long, slow process, but eventually, the monkey reaches the part of the forest with many bananas. This assumes, of course, that there are branches that are close to each other along the way. Hidalgo and Hausmann's breakthrough is that they show through sophisticated statistical techniques that there is a pathway of branches from one side of the forest to the other.

This viewpoint gives us a clear understanding of how economies developed both in the past and the present. Commercial societies in northern Italy, Flanders, Netherlands and England pioneered the innovations that created high-value-added industries. This was a long, slow process because they had no one to copy at first. There was a huge amount of learning by doing, with many mistakes along the way. As long as four of the Five Keys to Progress created the necessary preconditions for progress, they could keep experimenting and innovating for centuries.

Eventually, this led to the critical breakthrough of the Industrial Revolution in Britain. This largely involved the application of fossil fuels to the critical transportation, communication, agricultural and materials sectors. Industrial technologies overcame some, but not all, of the geographical constraints on progress.

Today developing nations face a different problem. All of the necessary technologies, skills and organizations have already been invented, so they only need to copy them rather than creating them from scratch. But because skills typically require a face-to-face transfer, it is hard for people in developing nations to learn everything they need to know. Worse, they face highly competitive industries in the richer nations that are vastly more productive. The one important advantage that developing nations have is cheaper labor.

The logic of this viewpoint leads to a clear economic learning strategy for developing nations:

1. Identify all your existing domestic industries.
2. Identify other industries that are both closely related to those industries and which have equal or higher added value.

3. Leverage the necessary technologies, skills, organizations and capital from existing industries to learn the new skills in the new industry and use the advantage of cheaper labor to outcompete richer nations. This will often involve selling an existing product at a cheaper price than richer nations can produce it for. It also involves focusing on the low end of the market where price is more important than quality while leaving richer nations to focus on the high end of the market.

4. Keep repeating the process for decades.

> **KEY INSIGHTS**
>
> To successfully export, a city needs the necessary technology, skills, organizations and capital. These factors are typically acquired by copying them from richer regions that already have them, modifying them for the local environment and then slowly learning by doing.

It is a nice theory, but does it work? Hidalgo, Hausmann and others give compelling evidence that economic complexity is a useful concept, but it is not yet clear whether it can be put into practice.

By statistical analysis, they show that complexity explains 73% of the variation in income across 128 countries. They also show that the difference between national income and level of complexity is the single best predictor of future economic growth. In other words, poor countries that have established a beachhead in higher-value-added products experience much higher growth rates in the immediate future than those who do not.

What economic complexity theorists have not proven is that poor and developing nations can use their theory to bootstrap their way up the value-added ladder. I know of no nation that has intentionally implemented strategies derived from the theory of economic complexity and then shown clear results in the form of higher economic growth. This is probably because the theory of economic complexity is new and relatively unknown to developing nations.

Nor have economic complexity theorists shown that rich nations actually followed something close to their recommendations in the past.

Of course, these nations could not possibly have been acquainted with modern economic theories, but the logic in their time would have been equally applicable. From my readings of economic history, I believe that Europe, the United States, Japan and other newly industrialized nations in Asia have functioned exactly as the theory predicts, but it is hard to present quantitative proof without good data.

We need economic complexity datasets to reach further back in history. Currently, there is little data before 1970. Adding more data might allow us to prove that cities and nations have gradually ratcheted their way up from low-value-added industries to related industries of higher value. They have done so by a blend of learning new skills, copying earlier innovations and local experimentation. Those nations then repeated the process in another related industry. Once one understands the logic of economic complexity, it is difficult to see how they could have grown any other way.

In the meantime, I believe that developing nations (and rich nations to a lesser extent) would be wise to tailor their national economic policies to the theory of economic complexity. It seems much more useful than the vast majority of the advice that other Western experts are currently giving them.

Key #5: Fossil Fuels

Before the Industrial Revolution, many Commercial societies in Northwestern Europe experienced progress. This progress created a standard of living that far surpassed other societies that had existed up until this time. But by today's standards, citizens of Commercial societies were still relatively poor.

The key missing ingredient was widespread use of fossil fuels, which enabled humans to acquire huge amounts of concentrated chemical energy and transform it into useful energy to perform tasks. Most importantly, that energy could be used to power new industrial technologies.

The Industrial Revolution in Britain added this fifth and final Key to Progress. The result was key technological innovations that transformed daily life. The railroad, steamship, steam turbines, automobiles, trucks, airplanes, electric motors, container ships and the electrical grid, are just a few of the thousands of industrial technologies that we take for granted today.

These fossil fuel-based technologies led to a standard of living for a typical person far beyond anything the richest men of the pre-Industrial era could imagine. Before the use of fossil fuels, economic growth and technological innovation mainly benefitted a very small portion of the world's population. Today, to a large extent because of fossil fuels, economic growth and technological innovation benefits the vast majority of the world's population.

Fossil fuels are critical to progress and economic growth because of their incredible energy density and the fact that they are affordable, easily stored and transported, reliable, controllable and easy to scale to fit needs. And because of these characteristics their geographical limitations are radically less than virtually all other energy sources.

All of these advantages explain why fossil fuels offered huge advantages over pre-Industrial energy sources such as human-power, animal-power, wind-power and water-power. Human-power and animal-power require food, which was the critical constraint on traditional societies. Humans and animals need food to survive and reproduce and they need far greater amounts of food to increase production beyond subsistence.

Before the Industrial Revolution, societies were caught in a "Catch-22" situation. You needed more energy and food to create progress, but the people and animals required to produce that energy required more energy to do so. This is why there was such a long delay between the invention of agriculture and the Industrial Revolution. Gradually, a few societies overcame those limits by increasing per capita food production and distributing the gains to productive cities.

In addition to being one of Five Keys to Progress, fossil fuels are also critical for two of the other keys: highly productive agriculture and export industries. Fossil-fuel-powered tractors, synthetic fertilizers and petroleum bi-products played a critical role in radically expanding agricultural productivity.

And while some industries do not require high amounts of fossil fuel usage beyond electricity, these are mainly industries dominated by rich nations. Manufacturing is critically dependent upon affordable, controllable energy that only fossil fuels can provide.

Quite simply the prosperous world that we live in today would not have been possible without the widespread usage of fossil fuels. The Industrial Revolution in Britain might have been fueled by imported wood and charcoal, but it surely could not have spread to the rest of the world and lasted for centuries without fossil fuels. Without industrial technologies powered by fossil fuels, most nations would still be living at the same standard of living as they did in 1500: i.e. desperate poverty for virtually everyone but a few elites and a few lonely commercial cities.

While we have developed other industrial energy sources since the Industrial Revolution (nuclear, hydroelectric, solar, wind, biomass, geothermal), none of them offer all or even most of the advantages of fossil fuels. In particular, all but nuclear have very serious geographical constraints and much lower energy density, while solar and wind are unavailable during large portions of the day and year.

Proponents of solar and wind are correct in that their prices are falling rapidly, but they do not acknowledge that price is only one of constraints on our ability to replace fossil fuels with an alternate energy source. Indeed, solar and wind require vast amounts of fossil fuels to construct in the first place.

One important point is that the use of fossil fuels is far more important than the production of it. Many nations have industrialized without domestic resources of fossil fuels, although domestic coal sources have been very important. Because of industrial transportation technologies and energy companies, anyone can now buy fossil fuels on the market.

And some nations have been seriously hurt by domestic fossil fuel production. While the United States, United Kingdom, Canada and Norway have largely benefitted from domestic production, a long list nations such as Saudi Arabia, Russia, Iran, Kuwait, Venezuela, and Mexico have been hurt by such production. Their economies are dominated by one industry, and their political elites largely control that industry.

KEY INSIGHTS

Today, to a large extent because of fossil fuels, economic growth and technological innovation benefit the vast majority of the world's population.

In general, nations which acquired four of the Five Keys to Progress *before* discovering coal, petroleum or natural gas have benefitted most from those discoveries. Those nations were effectively adding new revenue streams to an already dynamic and stable political and economic order based upon transparent, non-violent competition between elites.

Nations that discovered fossil fuels before they acquired four of the Five Keys to Progress have usually experienced the opposite. They rapidly evolve huge government monopolies to extract and distribute fossil fuels. Political elites then use those monopolies to distribute wealth to a small group of political and economic elites to maintain power. The last thing these elites want is transparent, non-violent competition between each other. Nor do they want other high-value-added industries that could potentially be used to finance political opposition.

These nations have effectively established a modernized version of what Agrarian elites have been doing for thousands of years: extracting wealth from society and using it for their own purposes. The enormous amounts of wealth generated by fossil fuels give elites a new lease on life and enable them to effectively ignore demands from the masses for more freedom and broad-based economic growth.

I do not claim that fossil fuels were the only energy source used by Industrial societies to power their progress in the past. Nuclear power and hydroelectric dams have provided a significant amount of electricity in the 20th Century, and they still do today. Unfortunately, their benefits are restricted to the production of electricity. This makes their application to industry and transportation far more limited than fossil fuels.

In most circumstances, nuclear and hydroelectric are more expensive than fossil fuels and require longer and more expensive time to construct. During construction, they also require substantial amounts of fossil fuels to power construction and transportation equipment, as well as the production of steel and other materials. Hydroelectric dams also require very specific geographic characteristics that most nations do not have in abundance.

Most importantly, neither energy source was important for any nation during or before their transition from poverty to progress. They only became important after they transitioned to progress. So widespread usage of fossil fuels is a Key to Progress, while nuclear power and hydroelectric dams are the results of progress. They are a useful supplement to fossil fuels, but they cannot entirely substitute for them, particularly in the realm of transportation and industry.

If you disagree on those points, you are free to substitute the words "fossil fuels" with "fossil fuels later supplemented by nuclear power and hydroelectric dams where geography and capital make them possible". In this book, I prefer the more concise version of just saying "fossil fuels" for obvious reasons.

Nor do I claim that we will never invent another energy source that has all the advantages of fossil fuels and none of their disadvantages. I believe that as long as progress is maintained for the next century, it is very likely, perhaps inevitable, that we will do so. In fact, in future books in this series, I suggest policies that can help make it happen sooner.

Nor do I believe that solar, wind and other non-hydro renewable energy sources cannot play a role in Industrial societies. Their use is rapidly increasing and their cost is rapidly declining. That is a good thing, and it is a result of the vast, decentralizing problem-solving network that is a modern society.

My claim is that solar and wind have never fully substituted for fossil fuels, nor can they do so within the next one to three decades. This is not the place to discuss all the positive and negative consequences of replacing fossil fuels with solar and wind, but I believe that climate activists are naïve in believing that we can make such a transition quickly using current technology without major consequences for economic growth and progress.

It is not a coincidence that the percentage of energy consumption made up of fossil fuels has hovered just above 80% for the last few decades. This is despite trillions of dollars of investments in renewable energy and constantly increasing global energy usage. Solar and wind are growing rapidly, but they are not replacing the use of fossil fuels to any significant extent. They are merely adding energy in addition to the current usage of fossil fuels. I believe that trend will continue for the next few decades.

Today virtually every mention of fossil fuels highlights the negative consequences of their use, particularly pollution and climate change. But it is important to realize that fossil fuels, despite their drawbacks, are a key foundation of progress. Quite simply, the modern world that we take for granted would not have been possible without fossil fuels.

Fossil fuels power innovation. Fossil fuels power economic growth. Fossil fuels power our education system, our transportation system, our communication system, our food production system, our health care system, and our military. Fossil fuels are key to generating all the wealth that pays for every government program we have. Before we try to eliminate fossil fuels, we need to make sure that we do not also eliminate all the benefits that have come from their use.

> **KEY INSIGHTS**
>
> Fossil fuels are critical to progress and economic growth because of their incredible energy density and the fact that they are affordable, easily stored and transported, reliable, controllable and easy to scale to fit needs. And because of these characteristics, their geographical limitations are radically less than virtually all other energy sources.
>
> Before we try to eliminate fossil fuels, we need to make sure that we do not also eliminate all the benefits that have come from their use.

What About All Those Other Theories?

To the best of my knowledge, no other author has identified these Five Keys to Progress. Many have written about some or all of them and noticed their importance, but then chosen to focus on other causes that they felt were more important. A few focus on one of the Five Keys to Progress and consider that one key to be the most important precondition but ignore the other four. To the best of my knowledge, none have focused on all five and shown why other potential causes are not as important as they seem.

One needs to be leery about announcing an intellectual breakthrough. The topic that I am addressing is one of the most controversial and heavily researched in all of history and social science. Dozens, if not hundreds, of books have been written on the subject. Among the causal factors that other authors have identified include freedom, property rights, the rule of law, free trade, patents, democracy, military competition, separation of church and state, railroads, education, literacy, Protestantism, Christianity, Western family structures, the fall of the Roman empire, the Enlightenment, useful knowledge, the scientific method, capital markets, communication technology, feudalism, manorialism, modern medicine, transportation, religious pluralism and many more.

And the above list is far from comprehensive. If I made a comprehensive list of all the presumed causes of progress in the literature, I could probably fill this entire page.

If you strongly agree with one or more of the above factors and find them more persuasive than my arguments, I would encourage you to carefully read *Bourgeois Dignity: Why Economics Can't Explain the Modern World* by Deirdre McCloskey. In her book, she thoroughly demolishes all of the other competing explanations for what she calls "The Great Enrichment" or what I call "progress."

I could not explain it all better than she did, so I decided to not even try. Plus not doing so has the bonus of cutting the length of this book in half by merely saying: "Read her book!" I am sure that my readers will be much relieved. More importantly, I am relieved that my book was published after hers, so she has not yet rebuked my work in public.

As an aside, I find McCloskey's theory that the cause of progress is the two ideas of Bourgeois Liberty and Dignity to be less compelling than her criticism of other theories. At its most fundamental level, I am a materialist and McCloskey is an idealist.

I would ask McCloskey: "So why did the idea of Bourgeois Liberty and Dignity arise when and where it did? And do you have evidence that it arose *before* the rise of Commercial cities in Northern Italy in the 1200s? If it came from elsewhere, did it come before or after the bourgeois came to play an important role in those societies? And if it came from elsewhere, why did it not trigger the Great Enrichment there?

I can find no logical reason why respect and admiration for the bourgeois would emerge in a society *before* the bourgeois started to play an important role in that society. Nor can I find any evidence that it did; granted, without polling data, it is difficult to know what the masses thought 800 years ago.

I believe that Bourgeois Liberty and Dignity came *after* the bourgeois showed results in the city/states of Northern Italy. So I believe that McCloskey has identified yet another important result of progress, not a cause of it.

It is the Five Keys to Progress that are the ultimate causes of progress. As we will see, this cause-or-effect problem is common to virtually all the other proposed causes in the current literature.

If you do not believe me, go through the list that I made above and ask yourself for each of them: "Do the theorists explain where these supposed causes came from in the first place? What caused the cause?" Very few have an adequate answer to that question. And even when they do, I would just keep asking the same question for each additional cause. It just pushes back the problem to an earlier time.

Because my theory includes a full causal chain that goes back to forces that existed before modern humans evolved — biology, geography, energy and evolution — it does not have the same problem as the other competing theories.

Why It Took Us So Long to Discover the Five Keys

Getting back to the list of proposed causes made be previous authors, its length should make one immediately wonder: if so many presumably intelligent writers who had a deep knowledge of history came to so many different conclusions, how will we ever know the real causes of progress? And more importantly, why should I think that I have got it right and everyone else got it wrong. Why has no one else identified the Five Keys to Progress?

Here is the problem: almost all of the previous authors are partly correct, but they often get the causality backwards and miss important feedback loops. They are all correct because each of the identified factors were *associated* with progress and long-term economic growth. They believed that these factors were causes of progress, but those factors were actually benefits of progress.

Some of those benefits then feed back into the process to create a positive feedback loop. That is why benefits of progress can often be mistaken for causes of progress (Ignore the section in this graphic labeled as "How Progress Works." I will discuss those factors in the next chapter).

The most immediate causes of ongoing progress are, in the words of social scientists, "over-determined." In other words, there are so many causes that appear to create progress, and they interact with each other in complex ways. So, it is almost impossible to separate their effects out to find that one key causal factor. To make it worse, we often have a critical lack of datasets that makes it difficult to apply sophisticated statistical techniques to untangle the causes.

Typically, previous researchers focused on a specific time period and geographical region, identified what seemed to be an important factor that separates a specific nation or region from others during that time period, and then examined how that factor developed over time.

Most researchers focused on one cause, but others focused on a handful. Some of them used sophisticated statistical techniques to isolate the effects of individual variables.

The problem is that these methods are very sensitive to the time period and region that the author selects. When one "freezes" progress at any one time and place, there are a huge number of factors that appear to have caused that progress.

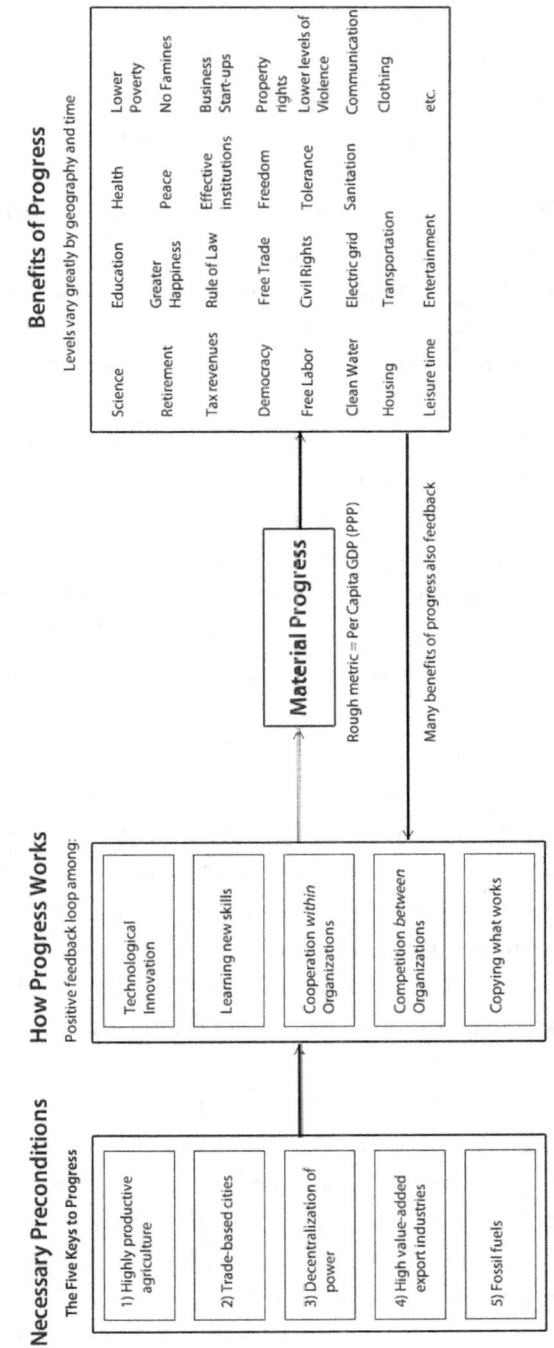

Understanding Progress

Necessary Preconditions
The Five Keys to Progress

1) Highly productive agriculture
2) Trade-based cities
3) Decentralization of power
4) High value-added export industries
5) Fossil fuels

How Progress Works
Positive feedback loop among:

Technological Innovation
Learning new skills
Cooperation *within* Organizations
Competition *between* Organizations
Copying what works

Material Progress
Rough metric = Per Capita GDP (PPP)

Many benefits of progress also feedback

Benefits of Progress
Levels vary greatly by geography and time

Science	Education	Health	Lower Poverty
Retirement	Greater Happiness	Peace	No Famines
Tax revenues	Rule of Law	Effective institutions	Business Start-ups
Democracy	Free Trade	Freedom	Property rights
Free Labor	Civil Rights	Tolerance	Lower levels of Violence
Clean Water	Electric grid	Sanitation	Communication
Housing	Transportation		Clothing
Leisure time	Entertainment		etc.

It is often difficult to separate cause from effect. It is also very difficult to separate proximate causes from ultimate causes. Indeed, there is a common, though unconscious, practice of completely neglecting necessary preconditions for progress in favor of focusing on more immediate factors.

I believe, however, that there is another way. Rather than looking at all the factors that a prosperous societies requires to continue its progress at any one specific time and place, we need to identify the few factors that are both necessary and sufficient to transition a society from poverty to progress in the first place. This makes it essential to identify when progress got started.

I view progress as an evolutionary process that was initiated by the establishment of certain preconditions: The Five Keys To Progress. We can even call those preconditions "ultimate causes." Progress can continue as long as those preconditions persist.

Just like biological life, another evolutionary process, a society that is experiencing progress is composed of individuals that will find solutions to get all the other things needed to survive, reproduce and expand. These solutions can be thought of as the proximate causes of progress at a specific time and place. But they are not necessary preconditions. They are, in fact, the result of those necessary preconditions being met. There are dozens, if not hundreds, of factors related to progress, but there are only five necessary preconditions.

Once a society acquires the five keys, society changes from a relatively stagnant and poor society to a vast, decentralized problem-solving network. That network is full of individuals who have both the ability and the incentive to acquire all the other factors that are necessary to keep progress going. It may take a great deal of time, but some of those individuals are guaranteed to succeed if the necessary preconditions persist.

For example, a society does not need much literacy to start the transition to progress, but once the transition occurs, the society has a strong incentive to found schools and adopt literacy as a part of their curriculum. Having a skilled workforce is key to further development, so support rapidly builds for a solution.

For the earliest societies to experience progress, this meant inventing an educational system or at least radically reforming an existing one. For later societies, they can merely copy what is already working in the richer societies that have experienced progress for longer and then modify those solutions to better fit local conditions.

For these reasons, literacy is a critically important benefit of progress that also helps to accelerate progress. But literacy is not an ultimate cause of progress. This logic can be applied to dozens of other factors that I list as benefits of progress in the preceding graphic.

Once society transforms into a vast, decentralized problem-solving network someone will find a way to either innovate or copy all the other factors that are needed to continue progress. In other words, previous researchers have largely focused on proximate causes and benefits of progress that will naturally evolve once progress initiates, whereas they have missed the Five Keys to Progress that are necessary preconditions. Progress cannot start or continue without significant progress on all of those five keys.

So I believe that once progress starts, as long as the preconditions are maintained, all the solutions required to expand progress will inevitably be found given enough time. If I am correct, this undermines the vast majority of the supposed causes of progress. This is why I see those supposed causes of progress as being the results of progress.

All of this means that most researchers came to incorrect conclusions and currently give incorrect policy advice. Worse, they do not realize that they are basing their policy advice on flawed assumptions, and they encourage others to come to the same wrong conclusions.

Or to put it another way, they are so focused on an individual tree that they cannot see the forest. We really need to understand that one individual tree or the species that this tree belongs to, but we cannot completely ignore the entire forest.

Why have so many previous researchers made this mistake? I believe it is because they have not had a wide enough timeframe, and they have not investigated enough societies or potential causes. Many investigators of progress and economic growth focus on this year or the last decade. A few more investigators focus on the last 50 years. Some go back as far as the Industrial Revolution and investigate the subsequent 200 years. Most modern investigators focus on Britain, the United States and Europe. Some focus on one nation or region outside the West.

Other researchers come from the opposite end and examine pre-Industrial history, but they do not apply what we know from more recent history to their findings. Researchers of the distant past (say more than 1,000 years ago) tend to focus on biology, culture, geography, religion, language, philosophy, state-building and wars. Researchers of today and the last 200 years tend to focus on technological innovation, science, institutions, and economics.

Very few researchers cover both pre-Industrial and Industrial history and focus on all the potential causal factors listed above. So they have not been able to find the right answer because they have looked in the wrong places and started with the wrong assumptions.

Another key part of the problem is the widespread belief that progress started with the Industrial Revolution in Britain sometime between 1750 and 1850. Most researchers start with an assumption that there was a pre-Industrial era and a very different Industrial era. Typically those researchers have only a basic knowledge of the pre-Industrial era and use the theories of Thomas Malthus to explain the time before the Industrial Revolution. They then focus on the incredible innovations of the Industrial Revolution and the following centuries.

This focus on the Industrial Revolution is useful, as it was one of the most impactful developments in world history. A huge number of books and articles have investigated the causes and effects of the Industrial Revolution.

The list of proposed causes is almost as long as the earlier list that I mentioned earlier. Though they often disagree, our knowledge of progress has been greatly advanced by these researchers. But all their work rests on assumptions that are partially flawed.

I believe that progress started centuries earlier, and it started because four of the Five Keys to Progress evolved during that time period in a few geographical areas. The reason why that earlier progress has been obscured is that it was restricted to a few small commercial city/states in Northern Italy and then Flanders. It then expanded to the relatively small nations of the Dutch Republic and Southeastern England. France, Spain, Russia and what later became Germany dwarfed these regions in size and population.

In addition, there are very few researchers that have focused on all four of those societies at once. There are a huge number of books about pre-Industrial England and a much smaller number of books on Northern Italy and Flanders in the Late Middle Ages and the Dutch Republic. Almost no books examine all four societies at the same time, and notice their strong resemblance with each other and their strong differences with other societies in Europe and Asia at the same time.

Unless a researcher focuses on those specific Commercial societies, it is easy to miss early progress. Even European historians of the period often miss it or underestimate its relevance to today.

In particular, many specialists of this time period underestimate the differences between the Commercial societies that invented progress and the Agrarian societies that dominated the politics, wars and population of Eurasia. Indeed, they do not even apply the essential terminology of Commercial and Agrarian as labels for these societies. They do so because they underestimate the essential differences between those two types of societies.

Previous progress in Northern Italy, the Low Countries and pre-Industrial England has often been hand-waved away as being merely temporary "efflorescences." This term brings to mind a beautiful flower that wilts after only a few days or weeks. But these efflorescences before the Industrial Revolution lasted centuries. Some of them lasted longer than the time period between now and the Industrial Revolution. They are no more temporary or short-term than the progress of today.

And these researchers ignore the fact that these periods of very real progress ended, not because of inevitable internal causes, but because of military invasion by more powerful kingdoms. If Nazi Germany had won WWII or the Soviet Union had won the Cold War and invaded Western Europe and the United States, killing progress and economic growth, we might consider the Industrial Revolution to be a temporary efflorescence as well. Military power can often destroy progress, but that does not mean that progress never existed.

The real reason that the Industrial Revolution was so critical was not that it created progress, but because it radically accelerated it and spread it to geographical regions where progress was not previously possible. Just as important, the Industrial Revolution radically changed the balance of military power between societies experiencing progress and those that did not. This lowered the threat of military conquest that had previously killed progress in militarily weaker societies.

Before the Industrial Revolution, progress was restricted to very small city/states or small nations. Those nations had relatively small populations. They had highly effective armies and navies, but they were typically smaller in size than their rivals.

Those societies were always under threat from larger and more populous Agrarian regimes. Those Agrarian regimes had powerful militaries that threatened the very existence of those city/states and, therefore, threatened the continuation of progress. The Industrial Revolution turned Britain and particularly the United States into powerful military machines that played a critical role in winning World War I, World War II and the Cold War.

Just as importantly, the Industrial Revolution created an entire suite of revolutionary technologies powered by fossil fuels that enabled progress to expand to much of Europe, North America and Japan. More recently, progress has expanded to most of East, Southeast and South Asia, where the bulk of the human population lives. Even Sub-Saharan Africa has experienced very real progress over the last generation. This trend has had an enormous effect on the material standard of living for the bulk of humanity. It is an achievement that cannot be underestimated.

But the incredible importance of the Industrial Revolution can blind us as to the ultimate causes of progress. It was during the time period between 1200 and 1830 that a few small societies in Northwest Europe gradually evolved four of the Five Keys to Progress. Studying those societies enables us to realize that much of what we view as causes of progress are actually the result of progress.

Why Are the Five Keys Useful?

History matters. Humans have a natural tendency to focus on factors that seem the most relevant today. It seems so obvious that current factors are more important than the distant past. But we need to be aware that this "recency bias," as psychologists called it, often causes us to miss highly important factors that are better understood from a historical perspective.

Ironically, by looking further back in the past, our thinking becomes more relevant to the present. Those of us who are interested in keeping progress going in the wealthy nations and triggering it in developing nations need to be careful not to be dazzled by the blinding light of the Industrial Revolution and bleeding-edge technology. Doing so can easily cause us to misunderstand causality and give poor policy advice.

If we focus exclusively on more recent proximate causes, we are unlikely to give useful policy advice, and this will undermine the credibility of progress researchers and policy advisors. By pushing back our time horizon from today to a broad sweep of history, we can get a much better understanding of the actual preconditions for progress that are very relevant today.

I believe that the Five Keys to Progress is a unifying concept for understanding progress. I also believe that this concept enables us to answer some of history's most difficult questions as well as provide policy solutions for today. Using this idea, it is much easier to understand:

- The historical origins of progress.
- Why progress took so long to get started.
- How and why progress spread to different societies geographically over time
- Why so many poor nations were left behind for so long
- What rich nations need to do to keep progress going

- What developing nations need to do today to experience progress
- Which forces today threaten future progress.

From a more theoretical perspective, the Five Keys to Progress are a critical concept because they enable us to better understand history. We can understand why it took so long to achieve progress. There were simply too many preconditions for them to occur randomly in many places. In particular, agricultural innovation is really difficult and risky as well as highly constrained by geography and short-term variations in weather.

Even once a society evolves a productive agricultural system (the first Key to Progress), that society still has not accomplished much of value to anyone but farmers. The heavy lifting of progress takes place in the cities (the second Key to Progress). And much of the increased food surplus gets eaten away by children and the next generation (a key insight of Thomas Malthus). Worse, elites create centralized institutions that extract wealth from farmers and cities for their own benefit.

Without agriculture (by far the most difficult challenge), none of the other keys matter very much as no one has enough food to eat. Without cities, innovation and copying of innovation take place at an extremely slow rate. Without food or cities, progress is impossible to imagine.

Without non-violent competition between elites (the third Key to Progress), elite extraction of surplus food stifles the growth of cities and undermines potential export industries. It is much more comfortable for the elite to extract than to compete against each other. Older Agrarian regimes and more modern totalitarian regimes have experienced important progress by copying for a while, but they all need some form of political and economic competition to keep it going.

A few small city/states experienced progress before the widespread use of fossil fuels (the fifth Key to Progress), but this was seriously constrained by lower energy density. And it is possible to imagine energy technologies that can keep progress going in the future. But the Industrial Revolution and current economic growth could not possibly keep going without widespread use of fossil fuels.

Using the Five Keys to Progress, we can also understand why progress originated in Northern Italy in the 1200s and why progress popped up in other similar locations. Specific geographical and political preconditions enabled four of the five keys to potentially evolve. These innovations were far from inevitable, which is why I consider them to be Keys to Progress.

The Five Keys also enable us to see why some nations today have had such difficulty creating long-term economic growth, while others have been able to suddenly transform within one generation. Economists and development experts typically advise developing nations to establish free trade, reduce corruption, establish the rule of law and build institutions. In addition, many environmentalists advise them to radically cut their use of fossil fuels at a time when they most need them.

While some of this is not entirely incorrect, I believe that this is not very helpful advice. Nations generally acquire free trade, lower corruption, the rule of law and "good" institutions *after* progress has been well established. Those factors can stabilize economic growth, but they cannot trigger it.

Many newly industrialized nations, as well as the ancestors of nations that are rich today have followed very different advice from what is currently the conventional wisdom. We should copy what worked and modify it for current conditions.

The Five Keys to Progress also help us to understand what wealthy, Western nations need to do to keep progress going. I believe that the foundations of progress in Western nations are being undermined by an ideological group-think mentality, over-centralization, over-regulation, bureaucratization of institutions, and policies that attempt to abolish fossil fuels, limit agriculture productivity and make our cities unaffordable and less livable. All of these current problems stem from a neglect of the Five Keys to Progress.

Why Economists Are (Partly) Wrong

My use of the concept of necessary preconditions is very similar to what a large number of free-market economists believe about capitalism.

They argue that if you establish the necessary preconditions of free trade, property rights, rule of law and "good" institutions, people will find a way to solve problems given enough time. Those economists are not interested in explaining every technology, organization or policy that goes along with economic growth. Most economists assumes that, given the right preconditions, someone will figure out a solution.

To explain where I differ from those economists, I would ask them the next logical question: Where do all of those factors come from? What creates free trade, property rights, the rule of law and "good" institutions? Why did they evolve in Northwest Europe? How do you transfer those things to other societies? It is not clear the field has answers to those questions.

I argue that these "proximal" causes need to be explained by deeper "ultimate" causes. The same goes for historians who research economic history. Without the Five Keys to Progress, economic growth could not happen, at least not at a level that would lead to a long-term increase in standard of living for the masses.

So should we dismiss the importance of these proximate causes and the results of progress? No, absolutely not. They are critical to understanding progress. But if we are focused on keeping progress going in wealthy nations and helping developing nations to transition from poverty to progress, we must focus relentlessly on necessary preconditions (i.e. the Five Keys to Progress) and then, only when we are sure that these cannot be improved, we should focus on more immediate factors. Even if something is inevitable in the long run, someone has to do the work.

Conclusion

To sum up, progress is an evolutionary process, not one managed by any individual or group. It is the outcome of a vast, decentralized problem-solving network that comes into being when a society acquires the Five Keys to Progress.

Humans shifted from poverty to progress by accidentally inventing the essential preconditions for progress:

1. A highly efficient food production and distribution system.
2. Trade-based cities packed with a large number of free citizens possessing a wide variety of skills.
3. Decentralized political, economic, religious and ideological power.
4. At least one high-value-added industry that exports to the rest of the world.
5. Widespread use of fossil fuels.

Each of the Five Keys is necessary for progress, but none are sufficient by themselves. It is only when a society acquires all Five Keys to Progress that it creates sufficient conditions for Industrial progress in the modern era.

HOW PROGRESS
WORKS

I n the first chapter, I made the case that we live in a world of progress, and that progress plays an important role in making all of our material lives better. In the previous chapter, I explained the necessary preconditions for a society to transition from a state of poverty to a state of progress and keep it going: The Five Keys to Progress.

The Five Keys to Progress provide the essential preconditions for creating and sustaining the vast, decentralized problem-solving network that generates progress. In later chapters, I will explain how previous societies acquired those Five Keys.

The question naturally arises: how does this vast, decentralized problem-solving network function? What are the principle mechanics that enable this network to deliver progress both today and in the past?

In this chapter, I will explain how progress works. I will focus on the lower-level mechanics of that vast, decentralized problem-solving network. A good way to think of it is that once the Five Keys to Progress create a critical mass for preconditions for the network to emerge, these lower-level factors are what enables the network to produce progress.

How Progress Works

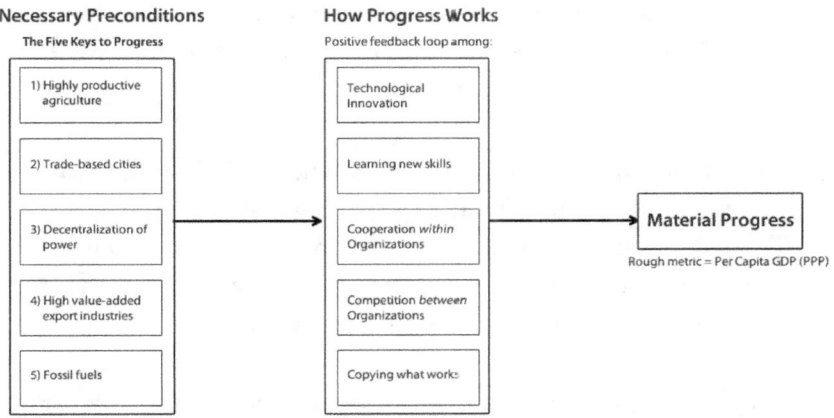

Under the right conditions, societies become a vast decentralized problem-solving network that generates progress. The network creates progress from a positive feedback loop among the following:

1. **Technological innovation**. This includes radical innovations such as the railroad, electrical grid, computers and the internet, as well as the ongoing incremental improvement and differentiation of thousands of other existing technologies.

2. **People learning new skills** to support those technologies. Without these skills, technologies are not useful, a fact that is often forgotten.

3. **People cooperating *within* organizations**. Those people work together using a wide variety of skills and technologies to accomplish a common goal.

4. **Competition *between* organizations** for scarce resources. In the past, this was usually food, while now it is usually revenue.

This competition forces organizations to embrace new technologies, skills and processes to out-compete other organizations. It also forces people within the group to cooperate more closely, and enables new organizations to be founded and older organizations to fail.

5. **People copying successful technologies, skills and organizations** and then modifying them to solve different problems. This enables innovations that work to spread into new companies, new sectors of the economy and new geographical regions. This step is critical to ensure that progress is widely shared.

6. **Consumption of vast amounts of useful energy**. Without energy, none of this can happen. Today the vast majority of that energy comes from fossil fuels.

The list above can be thought of as the most important factors in how progress works. For lack of a better term, I will borrow a term from biology and call them "behaviors." They are tasks that humans instinctively perform. Some of them are presumably encoded in our DNA, while others are conditioned by culture.

It is important to note that these behaviors are not the ultimate causes of what gets progress started. This is a mistake that many thinkers make, and it causes their advice to be unhelpful.

The Five Keys to Progress are the ultimate causes that create the initial conditions necessary for progress. This chapter explains in more detail specific human behaviors that make progress work while the Five Keys to Progress are in effect.

Looking Back At Our Metrics

Each of the metrics of progress examined in Chapter One involves the innovation of new technologies, the mastering of new skills required to build and use those technologies and the innovation of complex and diverse social organizations designed to most effectively apply those technologies and skills to solve a problem. These solutions invariably need some sort of energy source for them to work.

To pick one example, the metrics in Chapter One related to health (longevity, neonatal mortality, sanitation, drinking water, etc.) are driven to a large extent by technological innovations. These innovations include vaccinations, antibiotics, flush toilets, plumbing and sewage treatment systems.

Even simple medical technologies such as the medical thermometer, stethoscope, and hypodermic needles made it far easier for doctors and nurses to identify symptoms before treating diseases. More recent innovations, such as X-ray imaging, cardiac pacemakers, CT scanners, MRI scanners, prosthetics and pacemakers have made even greater contributions.

> **KEY INSIGHTS**
>
> Each of the metrics of progress examined in Chapter One involves the innovation of new technologies, the mastering of new skills required to build and use those technologies and the innovation of complex and diverse social organizations designed to most effectively apply those technologies and skills to solve a problem.

But each of these technologies requires skills. Each technology needs to be designed, produced, tested and distributed by companies that specialize in those fields. Just as importantly, doctors and nurses need to go through years of training to use them properly. Without those skills, these technologies would be no more useful than a rock lying on the ground.

In addition, social organizations had to evolve to coordinate all these skills and technologies. Corporations that manufacture medical devices had to be founded and staffed with people possessing the necessary skills. Hospitals had to be founded to coordinate all the doctors, nurses and administrators necessary to deliver diagnosis and care. Health insurance companies and government programs had to be founded to finance the payment for these services.

Those organizations in turn compete against each other for resources, usually in the form of revenue. They also compete for customers, investors and employees. Without this healthy competition, these organizations would be much less innovative. And they would not search as hard for "best practices" among their competitors.

And without vast amounts of energy, none of this would have been possible. Factories, hospitals, medical schools and other institutions require access to a stable, efficient electrical grid. The electricity in that grid is typically generated by fossil fuels, nuclear power or hydroelectric power. Fossil fuels are also necessary for the transportation of people and goods domestically and overseas.

Virtually all of these innovations started in either Northwest Europe or North America. This tiny portion of humanity has made the lion's share of innovations in technologies, skills and social organization.

In doing so, they pioneered solutions that are useful for all of humanity. These innovations were largely made to solve local problems, but once they were made, other people could see the results, purchase or copy the technologies, learn the skills and copy the social organizations.

Just as importantly, manufacturers figured out how to drive down the manufacturing costs for products that were at first luxury items. Only the wealthiest people could afford to buy these technologies when they were in their infancy. These wealthy customers gave companies a critical early market that paid for learning the skills and processes necessary by turning these expensive prototypes into affordable necessities. Corporations drove down the price enough so that poor people and poor nations could afford many of them.

When human beings live in the proper social and natural environment, they create societies that are vast problem-solving networks. When they live outside that environment, humans are only a little more innovative than chimpanzees.

Human history has been a long, drawn-out set of trial-and-error experimentations that has resulted in some human societies becoming vast problem-solving networks. This gives other societies the opportunity to copy them, though they often choose not to do so.

In modern societies we humans live in social environments where we can innovate technologies, learn skills related to those technologies and form social organizations that best utilize those technologies and skills. And we live in social environments that enable us to copy the innovations made by others.

Are These Behaviors Unique to Today?

It is important to note that all of these behaviors have been in existence since the advent of modern humans hundreds of thousands of years ago. Humans have always invented new technologies, learned new skills, cooperated in organizations, competed as groups against other organizations, copied other humans and consumed energy. It is quite likely that our hominid ancestors also performed behaviors that strongly resembled ours.

But until the Five Keys to Progress were acquired, the amount of change caused by these behaviors was so slow that they did not deliver progress — "the sustained improvement in the material standard of living of a large group of people over a long period of time." They delivered long, slow change, but no progress.

Before the Five Keys to Progress, human societies evolved without producing any progress for the masses. After the Five Keys to Progress came into being, human societies generated progress.

Even non-human animals can perform each of these behaviors to at least some degree. But no non-human animals perform each of these behaviors to the degree that modern humans do and with such a wide diversity of outcomes. The exception is, of course, that all animals consume energy because they must do so to survive.

Some non-human animals use tools and, presumably, those were "innovated" at some point. Beaver dams, bird nests, dolphins using sponges and chimpanzees using sticks are just a few examples. Only humans, however, have created a suite of technologies that is both vast in scope and enormous in diversity. Not surprisingly, some thinkers see this as what makes humans unique within the animal kingdom.

The same goes for skills. All non-human animals have skills (which biologists call "behaviors'). Presumably, those skills are largely encoded in their DNA, and in some species, the mother teaches additional skills. But no non-human animals have developed the enormous variety of skills that humans can perform. Juggling, ice skating, computer coding, writing books, archery, meditation, contemplating the meaning of life are just a few skills that some humans can perform. No other animal comes even close.

And the degree to which humans specialize in skills and get better at them is also unique. Some humans can juggle, but most cannot. Most people could probably learn to juggle, but they have chosen to specialize in other skills. It is not even clear that non-human animals can "choose" to specialize in a particular skill at all. Social insects specialize and many animals have gender roles, but these seem to be encoded in DNA.

All social animals cooperate in groups, but they all cooperate in groups that are virtually identical to each other. If you understand the dynamics of one pride of lions or one school of herring, you can presumably understand the dynamics of all prides of lions and all schools of herring. Non-human animals have almost no diversity in the type of groups in which they are associated.

Humans cooperate in families, bands, clans, tribes, nations, empires, corporations, labor unions, charities, schools, churches, hospitals, factories and numerous other types of organizations. Each of these organizations has different characteristics because humans voluntarily formed them to solve different problems. As far as I know, there is no analog in the animal kingdom.

In addition, a single individual human can cooperate in many different groups at the same time. I can cooperate within my family, my neighborhood, my metro, my nation, the company of my employer, my labor union, my favorite charity, and my church at different times. Modern humans naturally do this all the time without thinking much of it. This too has no known analog in the animal kingdom.

Because there are limited resources, these organizations are forced to compete against each other. The same is true of all territorial social animals. They compete ferociously to protect their territory and sometimes to expand into the territory of rival groups. Their competition is zero-sum.

Humans are also territorial and engage in violent competition. We call that competition "war."

But humans have also created an enormous variety of organizations that solve problems for people outside the group. Those organizations compete together non-violently, often in the marketplace. This non-violent competition encourages people to innovate technologies, learn skills, cooperate more closely within those organizations and create new organizations with different characteristics.

> **KEY INSIGHTS**
> Humans have a natural instinct to innovate, learn new skills, cooperate in groups, compete against other groups and copy each other. We need to consume energy to do so.
>
> But these behaviors only lead to progress, when The Five Keys to Progress are present.

So what do we take from all this? It is clear that these modern human behaviors that drive progress started to evolve long before even our hominid ancestors. This is an important finding because it links modern progress to biological evolution.

Humans have evolved behaviors that greatly increased both the magnitude and the diversity of non-human behaviors that have probably existed for tens of millions of years. It is only when these human behaviors are combined with The Five Keys to Progress that we see the emergence of modern progress.

So, enough about the animal world, let's get on to the modern human behaviors that drive progress.

Technologies Are Modular

As W. Brian Arthur points out in *The Nature of Technology*, all but the simplest technologies are combinations of less complex technologies. Take a bicycle, for example. A bicycle consists of many components: two wheels, a frame, the transmission, braking system and saddle.

Many of these components can be broken down further into

sub-components. The transmission component consists of a number of sub-components: pedal, crank arm, chainrings, cassette,
front and back derailleurs, chain and shifters.
The cassette sub-component consists of an
assembly of many gears.

> **KEY INSIGHTS**
>
> Most technologies are composed of many smaller technologies combined together.

The bicycle is a technology, but so are each of its components, as well as all of its sub-components and assemblies. Highly modified versions of these components are used in other forms of transportation technologies, such as motorcycles or dirt bikes. Slightly modified versions of these components are used in other types of bikes, such as commuter bikes, e-bikes, road bikes or mountain bikes.

Each of those bikes is related to each other because they have similar component technologies. But each uses different versions of those components to be more effective in solving different problems in different environments.

The motorcycle is more effective at carrying one person at speed on a paved road. The dirt bike is more effective at carrying one person at speed across a broken landscape. The mountain bike is more effective at giving one person aerobic exercise on unpaved surfaces. The road bike is more effective at giving one person aerobic exercise on paved roads.

Motorcycles and dirt bikes are very different from other bicycles because they have internal combustion engines. But internal combustion engines are components of a vast number of different transportation technologies: automobiles, trucks, railroads, and ships.

And the internal combustion engines themselves have a vast number of parts. Each type of engine has slightly different implementations of these parts to function optimally within that transportation device. Until one reaches the base materials that humans extract from nature, every technology is composed of other technologies.

The bicycle is an example of a synergy because the whole bicycle is fundamentally different from the sum of its parts. More importantly for humans, bicycles are useful pieces of technology — far more useful than each of its parts when separated. If one part of the bicycle falls off, it is suddenly less useful than it was when it was complete.

If the parts by themselves were more useful than the bicycle, they would never have been assembled in the first place. In addition, the bicycle as a whole can be improved by modifying one or more of its parts, leaving the opportunity for constant, incremental improvement. Smoother shifting transmissions can be created, along with lower resistance tires, more powerful brakes and a larger number of gears. Steel components can be swapped out in favor of titanium components to save weight.

Bill of Materials

All but the simplest technologies can be broken down similarly. To manage the full complexity of modern technology, industrial companies create a Bill of Materials (BOM) for all of their products. A BOM is a list of materials, sub-assemblies, parts and quantities that make up the product. Its purpose is to enable people to build, sell and service products exactly as they were designed.

Because technologies are modular, BOMs represent the product in a hierarchy of discrete parts. The finished product, for example, a bicycle is at the top level. All of its components are on the second level, for example, the wheels or transmission. The sub-components of the bicycle, for example, a cassette are on the third level: and we can continue all the way down to the base materials. For a complex technology, such as a commercial airliner, a BOM may include tens of thousands of components, sub-components, assemblies, sub-assemblies and parts, all the way from the fuselage to a lowly screw.

Bills of Material also define the final product in varying ways depending upon the occupation of the person viewing it. An Engineering BOM describes the final product as it is being designed or redesigned. A Manufacturing BOM describes how the final product is to be built. A Sale BOM describes how the final product is to be ordered from sub-contractors. Finally, a Service BOM describes the final product as it is to be serviced, diagnosed and repaired.

Even with all this complexity, BOMs often underestimate the

actual complexity of a piece of technology, as many companies buy assemblies or parts from other companies. Those companies have their own BOMs to describe all the parts needed to manufacture their "final" product.

A "BOM of all BOMs" that describes all the technology currently in existence as well as how to design, manufacture, distribute and service each of them would be of staggering size. Even a BOM for all the technologies of a typical Hunter-Gatherer society would fill an encyclopedia.

Innovation Is Recombinatory

Another observation made by W. Brian Arthur is that every innovation, no matter how different it appears from previous technology, is based on combinations of already-existing technologies. All ideas, no matter how profound, are based upon combining already-existing ideas. To innovate one needs to combine many existing technologies or ideas in a way that solves a problem more effectively. The whole is new, but the parts are copies.

When automotive engineers design a new car, they do not reinvent every part of the car. That would be extremely time-consuming and would sidestep all the learning that took place to optimize those parts by previous engineers.

Instead, engineers decide what problems need to be solved better. It might be the sale price; it might be fuel economy; it might be styling; it might be performance. Then they look for existing technologies that can be swapped out from existing models to solve those problems. It might be changing the tires; it might be adding cylinders to the engine; or it might be swapping out an internal combustion engine for an electric motor.

It is only when automotive engineers believe that they have used all possible existing parts that they experiment with building completely new parts. This is usually because all the other existing parts work perfectly fine, but there is one part that is a weak link. If they can drastically change that one part, then the new model can use the others parts more effectively. With one small change of a part, the new car can solve the chosen problem better.

Automotive engineers might decide to ask their tire sub-contractors to create a prototype for a tire using less sticky rubber. They might decide that building the frame out of aluminum instead of steel will save significant amounts of weight. They might decide to add in self-driving technology that is already being used by drones.

All of those changes can potentially lead to breakthroughs in automotive design. But everyone still recognizes the final result as a car. You can look at a Honda Accord today and a Model T built one century ago and immediately recognize that they are both cars. The Honda Accord is better in virtually every way than the Model T, but they are both still cars.

Honda engineers have achieved this goal by implementing all the design optimizations made for each of thousands of internal parts over a century. And every one of those parts is a copy that is either identical or closely resembles parts used in previous cars.

If one looks at the total amount of innovation in a current-model Honda Accord compared to all other cars, the changes are trivial. The Honda Accord, however, is a very impressive piece of technology because Honda engineers copied the innovations made by previous engineers and only copied those that worked best. Honda itself made some of those innovations in previous models, but other car companies or sub-contractors made the vast majority.

Even the most innovative companies in the world spend far more time copying than innovating. Or more accurately, they innovate by assembling unique combinations of existing technologies. They innovate by copying.

> **KEY INSIGHTS**
>
> Technologies can be recombined with each other to create new technologies.
>
> The more technologies there are, the more possible combinations exist. This makes the rate of innovation grow over time.

Technological Innovation Is Like Solving a Jigsaw Puzzle

An interesting analogy for technological innovation is trying to solve a jigsaw puzzle. Each piece of the puzzle represents an existing technology. The finished puzzle represents new technology.

Jigsaw puzzles typically come in boxes that contain all the necessary pieces for completing the puzzle, along with a photo of the completed puzzle on the cover. This photo gives the puzzle-solver important clues as to a general location where a specific piece might be located. Light blue pieces might be a portion of the sky, which is up near the top of the puzzle. Orange pieces might be the sun in the top right. Generally, one person sits down for an extended period to complete the puzzle, and when it is done that person might admire it for a while, but ultimately the final product goes back into the box.

The reason why a jigsaw puzzle is an apt analogy for technological innovation is that technology is combinatory and modular (as we saw earlier). Assembling a puzzle resembles combining separate pieces of technology to create a finished product.

Innovation is the opposite of breaking down technology into its components, sub-components and materials. Technological innovation is about assembling existing technologies to form something that has never existed before. Each component, sub-component and material already existed before the new technology existed, but together they make something unique, and hopefully useful. A new technology is, therefore, the first instance of a unique combination of two or more existing technologies.

But as we explore the analogy deeper, we notice that technological innovation involves a very different process to an individual solving a jigsaw puzzle. Innovation is different from a typical jigsaw puzzle in that:

Innovation is far more likely to be a made by a group of people than an individual. Despite history books being full of heroic individuals, such as Tesla, Edison, Ford and Jobs, innovation in the real world always involves many people, most of whom are not mentioned in the history books.

A jigsaw puzzle is not useful after it is finished. It is usually taken apart and then put back in its box. A new technology, however, is potentially useful if it performs a task better or more cheaply than existing technology. This means that it persists long after the "puzzle" is completed.

Even more importantly, a new technology becomes a potential component in future technologies. A new type of engine, software, hardware component or material can potentially be used in many technologies. Doing so creates the first instance of a new technology.

A puzzle has a front cover that helps people understand how all the pieces fit together. Technological innovation, however, has no equivalent. This makes creating new technology far harder than assembling a puzzle. People first must "design" how the pieces fit together before they can complete the puzzle. The design is an educated guess as to which pieces are necessary and roughly how they fit together.

A puzzle has exactly the right number and type of pieces. In technology, however, there are far more potential component technologies than are necessary to create a new technology. Some may appear to be useful in assembling the puzzle but are not. Others may look unimportant but are, in fact, key pieces of the puzzle.

In technological innovation, some of the "puzzle pieces" may be missing (for instance when a necessary component has not yet been invented). This missing piece then becomes its own puzzle that must be finished before the bigger puzzle can be completed.

If there is only one "missing piece", it may be relatively easy to conceive of and design it. This makes the primary problem one of implementation. But if there are multiple missing pieces, the "shape" of each of those pieces is unknown.

In other words, it may be difficult for people to even conceive of each of the component technologies that need to be invented. This makes it unlikely that the new technology can be created. Someone else may need to create a component, and then another team at a later date may try to resolve the problem. Because all the pieces are available on the second try, the "puzzle" can now be completed.

In puzzle building, each person knows about all of the puzzle pieces. In technological innovation, however, each person has an understanding of the proper use of only a few technologies. As the innovation of new technologies requires an understanding of numerous potential component technologies, this puts a serious limitation on the ability of one person working alone to create new technology. Therefore, a group with a diverse skill set has an advantage over an individual or a homogeneous group, whose members all have the same skill set.

People work on jigsaw puzzles in their free time, because it is fun. This limits the amount of time that most people devote to puzzle building. While many people find creating new technology fun, the vast majority who work on technology are paid to work full-time. This greatly increases the total number of man-hours devoted to innovating technology. Each man-hour devoted to solving a problem increases the chance that it will be solved.

Puzzle building is usually limited to one person or a set group of people. In technological innovation, however, people often realize that their group is missing a key skill but they know other individuals who possess that skill. Therefore, the more connected innovators are, the more opportunity they have to know a person who can help.

In puzzle building, the shape of each piece is fixed and it is considered cheating to change its shape to make it fit properly. In technological innovation, it is not unusual to modify a component technology so that it functions better within the new technology.

Doing so contributes to innovation in four ways. First, the modified component is effectively a new technology. Second, the new piece enables the higher technology to be created, effectively a second technology. Third, the modified component can be used to improve other technologies that were using the unmodified component. Finally, the modified component may turn out to be a key component that leads to one or more technological innovations.

Technologies Grow Exponentially

Yet another incisive observation by W. Brian Arthur is that, because any one piece of technology can potentially be combined with one or more other technologies to form a new technology, technological innovation tends to increase exponentially. This means that the rate of technological innovation is related to the number of unique pieces of technology that are already in existence.

We can see this with some hypothetical examples. If one society has 10 pieces of technology, that society will tend to have a far slower rate of innovation than another society that possesses 20 pieces of technology. The society with 10 pieces of technology will have 45 potential new combinations (9 + 8 + 7...) that potentially result in useful technology, while the society with twenty pieces of technology will have 190 potential new combinations. Most likely, the vast majority of random combinations will result in nothing useful, but the greater the number of potential combinations, the greater the likelihood that one combination will result in useful new technology.

The fact that technology compounds over time increases the advantages of the head start. Let's assume that 0.5 percent of all combinations result in useful technology every year. Over the long term, this leads to vastly different levels of technology between our two hypothetical societies.

After ten years, the society that started with 10 technologies acquires one new technology each year for a total of 19 technologies after a decade. The society that started with 20 technologies, however, would acquire 103 new technologies within a decade. One can easily see how small differences in the number of technologies can lead to vastly different technology bases at a later date.

The same goes for two different societies with slightly different success rates in creating new technologies. If the society that started with 20 technologies had a slightly higher success rate, let's say 0.6 percent, that society would have 288 total technologies ten years later. This is more than double the total number of technologies compared to the society with a success rate of 0.5 percent. Obviously, these numbers are completely made up, but they do show how subtle changes in the starting conditions can lead to vastly different outcomes at a later date.

Food & Energy

As Vaclav Smil has pointed out in his very impressive works on the subject, energy is critical to survival and innovation. The most important type of energy for animals is food. Food is essentially a form of energy that animals can consume to convert into useful energy to perform a behavior. Scientists call this process "respiration." The energy acquired from respiration is then devoted to behaviors that promote survival and reproduction.

Ongoing consumption of energy enables biological organisms to overcome the Second Law of Thermodynamics; they effectively create order from chaos (thus reversing what scientists call "entropy"). As soon as a biological organism loses the ability to consume energy and transform it to a useful form, it dies. The order of life is transformed back into the chaos of the rest of the universe.

For biological organisms, the ultimate source of energy is the sun. The sun fuses hydrogen into helium and releases huge amounts of solar energy. A small portion of that solar energy reaches Earth. Plants use the process of photosynthesis to consume that energy, then combine it with carbon dioxide and water to create sugars. Sugar enables plants to survive, grow and reproduce. Herbivores consume plants and use the energy to survive, grow and reproduce. Carnivores in turn consume the herbivores to fuel their survival and reproduction.

Human societies also rely on large amounts of energy to survive and reproduce. Just as with other biological organisms, the ultimate

> **KEY INSIGHTS**
>
> Energy is essential for life.
>
> For biological organisms, the ultimate source of energy is the sun.
>
> Human societies also rely on large amounts of energy to survive and reproduce. Just as with other biological organisms, the ultimate source of that energy is the sun. Energy in the form of food is the most important.

source of that energy is the sun. Energy in the form of food is the most important; without food, survival and reproduction are impossible. As we will later see, the quest to produce, prepare and consume food has been the dominant struggle in human history.

But humans have been able to innovate technologies that have enabled them to use other forms of energy: animal power, power from burning wood, windpower and waterpower. These additional energy sources have enabled humans to use far greater amounts of energy than any other species. Humans have used this additional energy to create far more complex technologies than would otherwise have been possible. In doing so, they have been able to solve problems far beyond the scope of any other animal.

The greater the technological base of a society, the larger the amounts of energy that the society needs to consume. As a society innovates new technologies, it sometimes identifies more efficient energy sources that can then be combined with natural materials to make useful technologies. These new energy sources effectively increase the rate of innovation.

Historically, by far the most important non-food energy source was fossil fuels. Fossil fuels have the critical advantage of being far denser (i.e. they have far more energy per unit of mass) than other energy sources used by humans. Energy density is critical to progress, as each additional unit of energy enables humans to innovate more specialized and complex technologies, skills and social organizations.

Energy and the Five Keys to Progress

Energy transfer also plays an important role in explaining why the five keys are so important to creating the necessary precondition for progress. Food, energy that humans are able to consume, is critical to the first key (an effective food production and distribution system). Without enough energy in the form of food, a society simply cannot grow very complex or create progress.

Fossil fuels, the fifth key, is also closely related to the concept of energy. Fossil fuels enable humans to capture the solar energy stored hundreds of millions of years ago and consume it today. This injects vast amounts of energy into society, making increasing complexity possible.

> **KEY INSIGHTS**
>
> Fossil fuels are ancient sunlight stored in the ground.

The other three keys, while not directly related to energy, can be understood as a form of energy transfer. Societies can grow complex and progress can evolve when the energy captured in the food surplus gets distributed to free people living in cities rather than elites bent on military conquest and conspicuous consumption. Export industries also transfer money and energy produced elsewhere and concentrate it in cities where it can be used most productively by society.

Technology Requires Skills

While biological organisms can survive and reproduce without human intervention (except for some domesticated plants and animals), technologies cannot exist without human intervention. Technology requires humans to possess highly specialized skills for it to survive and reproduce (by getting widely used by humans).

Even the simplest technology requires some amount of skill to use. In addition, conceiving of the technology, designing it, building it and repairing it are important related skills. Until people possess these skills, a specific piece of technology cannot come into being, or if it does, it would not last very long. It will certainly never spread far enough to become an important part of a society's technological suite.

Animals as a whole have an enormous variety of "skills" that humans cannot duplicate, but the repertoire of any one species is quite limited. Lions, for example, are very skilled at hunting large mammals along with other members of their pride. A large part of those skills are presumably encoded in their genes, while some come from copying their mother or from practice. But lions and other animals rarely learn a new skill that other members of their species do not possess. Eusocial insects, such as ants, termites and bees, have evolved a specialization that enables a broader repertoire of skills, but even those species are quite limited in their skill set in comparison to humans.

The collective skill set of even the simplest Hunter-Gatherer band, while very simple compared with other types of human societies, dwarfs the skill set of any non-human animal. As a society acquires more complex technologies, its collective skill set increases even more rapidly. This is because each technology requires a host of related skills. The more complex the technology, the greater the number of required skills.

While the total skill set of a society has no upper limit, there is an upper limit on the number of skills that one person can acquire. Learning a skill requires large amounts of time to practice and time is finite.

Cesar Hidalgo has developed the concept of a "personbyte" to denote the total amount of knowledge and skills that one person can possess. This concept is important because it shows that the only way for a society to increase the number of skills beyond a personbyte is for individuals to specialize in one skill or a small number of related skills. Fortunately, the more a person specializes in a skill, the greater the frequency of the repetitions and the more opportunity to get better at that skill. With an incentive to improve and actionable feedback, humans can become extraordinarily good at one skill.

The number of skills necessary to maintain just one technology in a modern economy is staggering. To get a glimpse, take a look at an organizational chart of any corporation or a jobs listing website. Among the broad categories of skills relevant to each specific technology are to:

> **KEY INSIGHTS**
>
> Every technology requires a large number of skills. Skills to use it, design it, test it, produce it, fix it and many more.
>
> So as the number of technologies grows, so must the number of skills grow even faster.
>
> There are only so many skills that one person can possess. So as the number of technologies grows, we must specialize. Fortunately, the more we specialize, the better that we get at each skill.

- Conceive of the technology (i.e. get an idea for a new product or service)
- Design the technology
- Test the technology
- Build the technology
- Market the technology
- Sell the technology
- Package the technology
- Ship the technology
- Distribute the technology
- Train end-users to use the technology
- Finance the technology (enabling customers to acquire the money to purchase the product)
- Maintain the technology
- Troubleshoot the technology
- Repair the technology

Most important of all, the customer/end-user must have the skills to use the technology. And the skills listed above are just broad categories. Within each category are specialized disciplines that are only known to the field.

To keep track of all the skills in modern society, the U.S. Bureau of Labor Statistics has created the Standard Occupational Classification (SOC). As of 2018, it consisted of 867 detailed occupations organized into 22 major groups. Even this very large number of occupations is just scratching the surface.

The official category for my profession — User Experience Designer — is "Web Developers and Digital Interface Designers." In practice, this category consists of dozens of specialties distinct enough that most employers would only look for a person within one of those specialties. And that is only one of 867 official occupations. Even within specialties, a typical worker will specialize in many skills known to only a small handful of employers.

This specialization creates a "Catch-22" situation for technological innovation. Technology cannot come into being and survive for any length of time without all the necessary human skills, but since the technology does not yet exist, neither do those necessary skills.

What this means in practice is that the process for a technology to go from an idea to a fully-realized product being manufactured and sold at scale will often take years or even decades. During this time period, humans must master the dozens of new skills required by the specific piece of technology.

In this way, innovation for an individual technology is a very slow process dominated by trial-and-error learning. But because in modern society there are millions of different types of technologies all evolving simultaneously, the sum total of all innovation is very rapid.

Social Organizations Knit Together Technologies and Skills

So far, we are left with the following: to achieve progress, we need technological innovation. To achieve technological innovation, we need skills. As the number and complexity of technologies accelerate, people need to specialize in a small number of skills. So technological innovation and progress fragment us into more specialized professions, each with clusters of related skills. But something is required to knit these specialized workers into a team that can produce a technology or service.

As shown by Robert Wright in *Nonzero*, that something is a social organization. Social organizations have existed throughout human history. The family, bands, tribes and nations are just some of the social organizations that humans have lived within. In modern times, social organizations have formalized into institutions, for example, governments, corporations, labor unions, churches, militaries, non-profits and many more.

For progress and technological innovation, corporations are the most important institution. Corporations knit together people with many different skills into one organization based on a business model designed to sell a small set of products or services to a specific customer base. Because each corporation has different technologies and different customer bases, each corporation evolves its own business model to succeed in that environment. Because customers have a limited amount of money, corporations are forced to compete with other corporations to survive.

> **KEY INSIGHTS**
>
> Social organizations enable humans with different skills to work together to accomplish a common task.

In this way, a corporation, like all social organizations, is much like a biological organism. While biological organisms compete for energy and nutrients, corporations and other institutions compete for revenue.

To survive this competition for revenue, corporations must adopt technologies, employ people with skills appropriate to those technologies and adopt processes that organize those people towards a common mission. Corporations that do this successfully will tend to acquire increased revenue. Those that fail to do so will tend to acquire less revenue. The worst will go bankrupt.

As technologies become increasingly more complex, requiring a greater number of specialized skills, corporations and other institutions must also become more specialized and complex. Traditional societies have a relatively small suite of technology, so they require only a few small organizations. Modern societies, however, have an enormous suite of technologies. This requires a vast number of complex social organizations, each specialized in narrow domains of technologies or services.

Prices

The mechanism that links these individual social organizations into a much larger problem-solving network is prices. Prices give incentives and they communicate information. If the price of an item or service is high enough that businesses can produce more of it while earning a profit, this communicates that the good or service is scarce. It also gives a strong incentive for an organization to form to deliver that good or service. If the price of an item is so low that those who produce it can barely make a profit, this discourages organizations from doing so.

A key benefit of prices and markets is that *they shift the focus from solving one's own problems to solving other people's problems.* Markets are often accused of encouraging selfish behavior, but they strongly encourage actions that benefit others.

Humans in modern societies solve their own problems, but, surprisingly, they do so by solving other people's problems first. Prices communicate which problems other people think need to be solved. People will pay more for a good or service that solves a more serious problem, so prices encourage them to solve their own problem (putting food on the table) by helping other people solve their problems.

> **KEY INSIGHTS**
>
> Prices shift the focus from solving one's own problems to solving other people's problems. Prices enable strangers to cooperate and give them an incentive to do so.

Of course, what I am describing is a classic free market as described by economists. But free markets, in themselves, cannot deliver progress. It is only with the Five Keys to Progress that markets can do so.

Markets existed during Hunter-Gatherer times. While they undoubtedly enabled people to trade for scarce raw materials like obsidian, there is no evidence that the result was anything resembling progress. If free markets had suddenly popped into existence in the 14th Century Incan empire, or the 10th Century Carolingian empire or in the 20th Century Congo or Somalia, there would have been no progress as a result.

Prices and markets are mechanisms by which the Five Keys to Progress enable cooperation over long distances, but they cannot deliver progress alone. However, as soon as the five keys join together in one society, markets with their pricing mechanism can enable progress.

Positive Feedback Loop

We have seen how technological evolution drives the acquisition of more specialized skills, which in turn drives the evolution of more specialized and complex social organizations. So just as biological evolution tends to create a greater number of species in a complex environment, cultural evolution and progress has the same effect on human societies.

But within this process, there are feedback loops that make progress even more dynamic. Humans do not just passively learn skills related to new technologies. They also push back to make the technology easier to use.

In my profession as a User Experience Designer in the technology field, I design the software that software developers write the code for. A key part of my job is to make complex technologies easy to use.

Early in the history of digital technology, business leaders expected users to be able to use technologies as they were provided. As long as a specific piece of technology was functional and performant, they thought the technologies would sell. As a result, early software was very difficult for novices to use.

Gradually customers and businesses pushed back on software companies and demanded that their products be made easier to use. More often, they just stopped buying the product and the software companies wondered why.

A few software companies, particularly Apple Computer, realized that software products needed to be designed for ease of use. In addition, technologies were invented to make user experience design more feasible at scale. The iPhone, Cascading Style Sheets and the Webkit browser engine played a particularly important role.

The combination of new technologies and design skills gave birth to my profession. The result is that the software market broadened beyond experts and gradually computers became something that users wanted to use in their homes.

And it was not just individual consumers who forced the software industry to adapt. Corporate consumers of software demanded that the product mesh with their overall business processes. Corporations were willing to invest in improving the productivity of individual departments, but if the output fundamentally disrupted the productivity of a dozen other departments, the cost was too great. So, just as software companies were forced to adapt to the skills of typical users, they were also forced to adapt to the processes of social organizations.

So while technological innovation forces society to adopt new skills and social organizations, the people with those skills and the people within those social organizations also forced additional rounds of innovation. The result is a feedback loop where innovations drive more innovations.

When new corporations are created to produce and use new technologies, those organizations put enormous pressure on existing corporations that produced or use older competing technologies. In many cases, these older companies were titans of industry that looked like they would dominate their sector for generations to come. But when a new technology evolves, seemingly dominant corporations can quickly be put out of business or at least radically diminished in their market share.

In theory, dominant corporations can simply adopt the new technology, but in practice, the skills and processes of the older technology may be so different from the skills and processes of the new technology that it is not easily done. Just like animals, big organizations built to survive in a certain environment cannot quickly adapt to major changes in that environment.

Changing the corporation so that it can produce or use new technologies may force widespread cultural changes that are strongly resisted from within. Typically, the established corporation is making enough money on the old technology that they do not even think that radical change is necessary.

The existential threat of the new technology does not become obvious until after revenues start to significantly decline. Just as older corporations need additional revenues to make the change, their revenues decline significantly. This often forces the once dominant company into a death spiral that leads to bankruptcy.

When an important new technology is innovated, it creates not just a few companies but an entire industry. In some cases, the new technology spawns dozens of new industries, each dedicated to building the components and sub-components of the new technology.

The invention of the automobile is a classic example. The automobile industry grew from a gaggle of small car companies to a vast sector of the economy with hundreds of large, medium and small companies building tires, brakes, steel, glass, along with a legion of specialized repair and parts shops.

Cooperative Competition

As societies get more complex, they spawn a greater number of more specialized social organizations. Because even the wealthiest societies have limited resources, each of those organizations competes against each other for those resources. And societies that inhabit the same region typically competed against each other militarily for survival.

As Peter Turchin argues in his book *Ultra Society*, this competition between social organizations increases the level of cooperation within each of the groups. We tend to think of cooperation and competition as opposites, but the two concepts are closely related. Competition *between* organizations increases cooperation *within* organizations.

I call this phenomenon "Cooperative Competition" (Turchin and biologists use the term "Multilevel Selection"). It is most easily seen in team sports. A team that has practiced and played together for years has a distinct advantage over another team that has never practiced together before, even if the latter is full of superior athletes. The more experienced team has worked out all the subtleties in meshing together skills and communication that enable teamwork. It is the same for any organization.

Being involved in a continual "us vs. them" struggle also fosters a culture that tends to contain the desire of individual members to free ride on the efforts of others within the group. Humans working within groups in competition against other groups have an amazing ability to subordinate their desires in favor of the needs of the group. They do so for the very reason that the survival of the individual is intrinsically tied to the survival of the group.

Competition between social organizations also fosters innovation. To gain an advantage over their competitors, social organizations are much quicker to innovate new technologies or copy those innovated by others. Organizations are much more likely to hire skilled workers or encourage their existing workers to learn new skills. Finally, they are also more willing to change internal processes to better utilize those new technologies and skills.

The very same organizations without competition behave very differently because their survival is not at stake. Rather than investing in new technologies, skills or processes, monopoly organizations tend to focus on creating a stable flow of resources from the outside, whether it is via higher prices, government subsidies or increased extraction.

For this reason, a society with a few dominant monopoly institutions is far less innovative than a society with many institutions that are competing non-violently against each other for survival. Monopolies stifle innovation even if they do not intend to do so. Institutions that compete non-violently against other institutions are forced to be innovative even if they would prefer to be monopolies.

Societies that allow new social organizations to be created and old social organizations to die enable innovation to take place. This innovation leads to progress. Societies that protect entrenched monopolies and interfere with the creation of new organizations that compete with those monopolies stifle innovation and undermine progress.

KEY INSIGHTS

Because there are limited resources, organizations are forced to compete against each other for survival. In the past, they competed over food. Now they compete for money.

Competition *between* organizations increases cooperation *within* organizations.

Competition between social organizations also fosters innovation and the desire to copy competitors to improve.

Large, Connected Populations Increase Innovation

As Matt Ridley argues in his book *The Rational Optimist*, one of the best ways for a society to increase its chances of successful innovation is to have a large population. Each individual added to a population increases the chances that someone will come up with a new idea. That new idea will be a unique combination of already existing ideas or technologies. The idea will be new, but the parts will be copies.

Think of each idea made by a person in society as a lottery ticket. The chances of success for each lottery ticket are low, so the best way to win is to purchase many lottery tickets. For this reason, societies with large populations tend to be more innovative than those with small populations.

Another way for a society to increase its chances of success is to be connected to other societies and be open to copying their ideas. No matter how populous a society is, the range of new ideas that its members can come up with will be constrained by cultural assumptions, geographic constraints, a group-think mentality and power politics.

Ten different societies with populations of one million that are connected via trade and migration are more likely to innovate than an isolated society with a population of ten million. The same number of ideas will be generated, but the difference between them will be greater, making it more likely that one will be an important breakthrough. The other societies will then have the opportunity to copy that innovation, but only if they are open to copying people from other cultures.

Crossing the Threshold to Progress

At some point, the total technology base accumulates so that society can start solving problems other than just acquiring food. People can start trying to solve problems related to health, education, transportation, communication, energy, and sanitation. Crucially, entire sections of society can start to focus on those problems, while other people can focus on acquiring and distributing food.

This helps us to understand how progress evolved out of apparent stagnation. When humans first evolved from other hominids, they had a very small technology base. Because the technology base was so small, the opportunities for innovation were also very small. Very little change could occur within each century. Compared to other animals, humans were innovating at a rapid rate, but the actual number of new technologies was still very small.

Despite this very slow rate of innovation, the total number of technologies gradually built upon each other exponentially until a food surplus was created. A single family could produce a bit more food than they needed to survive and reproduce. This food surplus increased the population substantially.

As we will see in the following chapters, however, this small food surplus did not lead to progress for the masses. Instead, it led to a zero-sum struggle for the benefits of that growth. The result was more babies and a few elites extracting the surplus for their benefit. Sadly, the very slow growth rate in society meant that it was easier to improve one's life by taking from others than it was to work and innovate.

But at some later point in time, a second threshold was passed into a state of progress. As the base of technology gradually expanded, creating ever more possibilities for new technologies, a few societies passed a point when it became easier to get wealthy by working and innovating rather than by taking from other people.

At that point, technological innovation became real progress that benefited the majority of people within those societies. That is the fundamental driving process that we saw expressed in the graphs in Chapter One.

KEY INSIGHTS

Societies that allow new social organizations to be created and old social organizations to die enable innovation to take place. This innovation leads to progress.

Large populations increase the chances of successful innovation.

Being connected to other societies and open to copying them helps good ideas to spread.

Innovation Is Local and Unequal

Innovations are local in nature. By this, I mean that the solutions evolve to solve local problems within their unique environment. Those solutions are not designed to solve all problems for all of humanity. Those solutions may be highly applicable to others, but that is not their original intention.

In addition, because the first instance of a solution is by definition located in one specific geographical area, it does not automatically spread to the rest of the world. The result is that evolutionary processes are inherently unequal.

An innovation made in Amsterdam, for example, does not automatically spread to the rest of the planet. So for at least some portion of time, Amsterdam (or at least a few people within Amsterdam) has technology that no one else does.

Because a larger number of technologies leads to a greater number of innovations, this process feeds upon itself. Areas that have local environments conducive to innovation create

> **KEY INSIGHTS**
>
> Innovations always start in one city and often take a long time to spread. This leads to inequalities between people.
>
> This enables highly innovative cities to "pull ahead" of the rest of the world... at least until they can copy the innovators.

more technologies, which make additional innovations possible. For this reason, in any given period there tends to be a very small number of centers of innovation that give birth to the lion's share of total innovations.

This means that, as long as those areas can maintain those characteristics that enable technological innovation, the innovation keeps feeding upon itself. This increasingly differentiates those few areas from the other human societies that have evolved in regions that were less conducive to innovation.

Innovations Can Be Copied

If those were the only factors that influenced innovations, then we would be stuck with a few very wealthy and innovative areas and the rest of the world would be poor and stagnant. Fortunately, there is one characteristic of human behavior that enables the innovations of a few centers to spread more broadly; it is the ability of humans to copy other humans.

Humans can see a technology or skill being used by another person, identify its purpose, assess its usefulness compared to their own toolkit and then decide to copy it. While other animals may have this skill (young birds and mammals learn from their mother), the human species has taken this to a new level. This means that humans from some societies can enjoy the benefits of innovations created by other societies.

There are real benefits to being the copier rather than the innovator. The innovator has to go through all the hard work of designing a new piece of technology, testing it and iterating on the results. The innovator also has to go through the hard work of learning all the necessary new skills related to that technology. They have also invested the time to learn how to adapt their social organizations on how best to use those technologies. The copier can just wait, watch the results and copy only when it looks like the original innovators have come up with a superior solution.

> **KEY INSIGHTS**
>
> The conditions that are necessary to copy technology are very similar to the conditions that are necessary to innovate technology. If a society has the proper conditions, it will be able to innovate and copy the innovations of others at a far more rapid rate than other societies.

Copying Is Not So Easy

Unfortunately, copying is not as easy as it sounds. To copy a new technology, a society must have certain characteristics. It is very difficult to learn a new skill entirely from scratch. It is much easier if one already has skills that are related to the new skills.

For example, it is much easier for a fisherman to learn skills related to fishing. A fisherman who is skilled in fishing with a hook and line would probably be able to learn how to fish with nets fairly quickly. The two skills are fairly closely related.

But it would be very difficult for that fisherman to learn quality assurance testing of software, as the two skills have very little in common. Without being able to interact directly with experienced QA testers, it seems highly unlikely that a fisherman would ever learn the skill with sufficient productivity to become an effective worker.

And even if this was a very talented fisherman who somehow learned the intricacies of software testing, what good would it do?

Without being in close geographical proximity to other people with the other skills related to software technology, his new skill would not do him much good.

Someone else in the village would need to learn how to code software, another would need to learn to market that software and another would need to sell that software. As each necessary new skill is added, the odds that a random group of people can learn them rapidly diminish.

Now, let us just assume that this is an extraordinarily talented group of fishermen. They somehow teach themselves all of the skills necessary to produce and sell a new piece of digital technology. It is very unlikely that the fishing village has a software company that could hire them.

In theory, all of the fishermen could band together to form their own company, but then they would have to learn all the processes (the ways that persons of differing skills cooperate to produce and sell a technology) on their own. They would also need to acquire all the technologies required to build and test software. Plus they would have to acquire the relationships with customers that companies typically possess.

Now, let us assume that our group of people who want to learn is instead an existing software company. The software company learns about a new type of software that a rival company has created. The software company will still have much to learn to effectively copy their rivals' products, but because they already possess technologies, skills and processes for similar software, they have a real chance of being able to copy with much less investment of time and resources than the original innovator.

This leads to an important insight: the conditions that are necessary to copy technology are very similar to the conditions that are necessary to innovate technology. If a society has the proper conditions, it will be able to innovate and copy the innovations of others at a far more rapid rate than other societies.

If the rate of innovation is fast enough that people see the benefits of working, innovating and copying the innovations of others rather than seeking to extract wealth from others via military conquest or expropriations, then progress can result. As long as these conditions remain in effect, it becomes a self-sustaining process.

Copy Your Way to Success

KEY INSIGHTS
No matter what occupation or hobby you seek to enter, there is one Golden Rule of Success: Copy the Successful.

The concept of innovation has been all the rage in 21st Century America. No matter where one looks — books, media or corporate board rooms — there is a focus on this concept. And for good reason. Innovation is critical to progress.

I would like to make the claim, however, that copying is at least as important as innovating. This is particularly true for those lower-income or young people who are just starting out.

No matter what domain you seek to enter, there is one Golden Rule of Success: Copy the Successful. It does not matter whether you want to be an athlete, artist, business leader, political leader, doctor, farmer or welder. The simplest way to get better at something is to copy those who are most successful in the field. By copying the best, you are effectively learning from all the trial-and-error attempts made by those who came before you.

In a fascinating article entitled "Why Copy Others?" a research team documented a computer simulation designed to test the most effective learning strategies. The researchers asked experts to write algorithms that would compete against each other in a computer tournament that consisted of 10,000 rounds. The entries consisted of various blends of learning strategies, including trial-and-error learning, copying and executing on a decision.

At the end of each round, successful algorithms "lived" (i.e. they went onto the next round), while unsuccessful algorithms "died" (they were dropped from the tournament). In this way, the researchers created a simulated evolutionary process with each player using different strategies.

So which strategy won? Those that relied heavily on copying what had worked in the previous round (or what the researchers called "social learning") repeatedly won.

The researchers did not expect this result. No matter how many times they ran the simulation, they got roughly the same results. The variations in outcome between the algorithms were mainly dependent upon the length of time spent learning from others and the timing of when enough information had been gathered before actually making a decision.

Additional experiments by Joseph Henrich (who has made stellar contributions to the field) show that copying multiple people is better than copying one person. Henrich and his assistant asked inexperienced students to recreate an image using image-editing software. Those students then passed on their image and instructions to the next set of students. Some of the students received the image and instructions from one student in the previous round; other students received the same from five different students.

The results were compelling. Students who received instructions from only one person hardly improved their skills at all. Among students who received instructions from five different students, however, the average skill increased dramatically (by 20-80%). Every student who received instructions from five people outperformed even the best student who received instructions from only one person.

This makes sense. As Henrich speculates, students who received instructions from five others were not just mindlessly copying. They were comparing the instructions from each of the five, looking for commonalities among those who were most successful. In addition, they were paying particular attention to the instructions written by the student who had done the best in the previous round.

By copying five people instead of one, they were effectively recombining the lessons learned from five different sources to create a new strategy that they could then try in the next round. The recombinations that succeeded best in the subsequent round were then given the most weight by the next round of students. So in each round students inherited the learning from students who had participated in earlier rounds.

Just as innovators recombine successful simple technologies, skills and organizations to create more complex technologies, skills and organization, copiers do the same. As one delves into the concepts more deeply, the line between innovating and copying becomes increasingly blurred. Innovation requires copying, and copying involves at least some innovation (mainly by making decisions as to which ideas to copy).

KEY INSIGHTS

Ideally copy a wide variety of successful people, and then figure out what they are doing in common. When in doubt, err towards copying the most successful person.

This leads to a valuable lesson for us all. The fastest and most effective means of success is to copy the successful. Ideally copy a wide variety of successful people, and then figure out what they are doing in common. When in doubt, err towards copying the most successful person.

Then once you have become one of those successful people in your field, you can start focusing on innovating something new. Only once you have exhausted all the lessons learned by previous generations of successful people is it worth spending the time and energy to innovate.

To summarize, innovating and copying the most successful innovations are critical factors in promoting progress.

Solving Today's Problems

An entrepreneur or engineer might say to me, "That is all very interesting, but I have to solve real everyday problems to keep my job. How do your ideas help me solve my customer's problems?" My answer would be to suggest that they apply the concepts from this chapter to their daily lives.

These concepts won't directly solve the immediate problem for the entrepreneur or engineer, but they might help them to think about solving their problems differently.

Most likely, the solution to their problem involves the application of technology, learning a new skill, implementing a new process, or creating a new organization. And what keeps those people focused on that problem is that they are in non-violent competition with other companies who are trying to solve the same problem to get money from customers.

Most likely, their solution will involve copying what a competitor does. If competitors do not have an immediate solution, the solution is likely to come from some sort of recombination of existing technologies, skills, or processes.

If your profession has been working on a particularly stubborn problem for a long time, perhaps you should look to other fields for inspiration. In *Range: Why Generalists Triumph in a Specialized World*, Daniel Epstein makes the compelling claim that many of the biggest intellectual breakthroughs come from analogical thinking.

Analogical thinking is where you shift the focus from solving the immediate problem to thinking about the high-level characteristics of the problem and then searching for solutions in other fields that solve related problems. It is reasoning by analogy. It is an extremely powerful tool that takes practice, but it will make your peers say: "Gee, that person is very thoughtful and innovative."

In today's hyper-specialized world, most professionals focus relentlessly on acquiring deeper and deeper knowledge about an increasingly narrow subject. Specialists are often so focused on solutions that they forget to fully analyze the problem at hand.

Being more aware of the bigger forces that are driving progress might just help give a solution to your problem. And by solving that problem, you are actually helping to drive progress forward!

Conclusion

Under the right conditions, societies become vast problem-solving networks that generate progress. That progress comes from a positive feedback loop among the following:

1. Technological innovation.
2. People learning new skills to support those technologies.
3. People cooperating *within* organizations.
4. Competition *between* organizations for scarce resources.
5. People copying successful technologies, skills and organizations and then modifying them to solve different problems.
6. Vast amounts of useful energy being injected into the system.

As technology becomes more complex (in that it has more components), it requires a greater number of skills to exist. Because a human being can master only a limited number of skills, this means that a greater number of people are required to cooperate. As the number of people and the diversity of skills increase, social organizations are required to knit these people together to solve common problems. Because each social organization must specialize in a specific type of technology, as the number of technologies increases, so does the number and complexity of social organizations.

No matter how many resources exist in a society, they are always limited in scope. This forces social organizations to compete against each other for survival. In traditional societies, social organizations competed for food. In modern societies, social organizations compete for money. To compete effectively, social organizations must embrace new technologies, encourage their people to master skills related to those technologies, and modify internal processes to best make use of those technologies. They must also build a common culture that enables internal cooperation.

All of these interacting factors of technology, skills, social organization, competition and cooperation create a positive feedback loop where innovation is exponential. The rate of innovation increases with:

- A large base of technology (which gives more technologies that can be recombined)
- A large amount of energy (from food and fossil fuels)
- A large population size
- Connected populations

In societies with a limited number of technologies, low amounts of energy and small isolated populations, the total amount of change is imperceptible even over the course of centuries. In societies with a large technology base, large amounts of energy and large populations connected with other similar societies, the total amount of change is significantly greater.

PROGRESS IS AN EVOLUTIONARY PROCESS

Throughout most of history, virtually all humans lived in social environments that were not very conducive to innovation. Only recently have societies evolved that maximize the possibilities of innovations that promote progress. Modern societies are problem-solving networks.

This evolution of modern societies that solve problems on a vast scale is, I believe, the most recent portion of an evolutionary process that goes back billions of years. This evolutionary process started from very simple sub-atomic particles immediately after the Big Bang, progressed to the formation of life on planet Earth and finally led to the emergence of complex human societies.

The evolutionary process shares many common characteristics with technological innovation. At its most fundamental level, this evolutionary process is about combining simple objects into unique combinations to create a new more complex object. This new object is a synergy that is fundamentally different from its components.

This new and more complex object can then be combined with other objects to create even more complex objects. Continuing this process over very long time spans created a complex physical world, a more complex biological world and an astoundingly complex human world.

Levels of Evolution

As Peter Corning argues in his book "Nature's Magic," this combination of simple objects to create more complex objects that are fundamentally different from their constituent parts is common to all evolutionary processes. One can think of the history of the universe as the development of ever more complex evolutionary processes.

These evolutionary processes can be broken down into four basic types:

1. Physical/Chemical evolution
2. Biological evolution
3. Cultural evolution
4. Progress

I will describe each of these briefly in the following sections.

Physical/Chemical Evolution

Physical/Chemical evolution started with the Big Bang billions of years ago. Because of the physical laws that came into being with the Big Bang, matter was transformed from simple particles into increasingly complex particles.

Major steps in Physical/Chemical evolution include:

1. Quarks combining to form hadrons. Each unique combination of quarks forms a different type of hadron. The most common types of hadrons are protons and neutrons.
2. Protons, neutrons and electrons combining to form atoms. Each unique combination of sub-atomic particles forms a

different type of element, each with unique properties.

3. Atoms combining to form molecules and compounds. Each unique combination of atoms forms a different type of molecule or compound. Each type of molecule or compound has its own unique properties.

The creation of more complex objects came about due to the interactions between the fundamental physical forces of:

1. The strong nuclear force (that binds the fundamental particles of matter together)
2. Electromagnetic force (the attraction/repulsion between charged particles)
3. The weak nuclear force (interactions that change particles)
4. Gravity (the attraction between objects that have mass)

Each of these more complex objects created by the interaction between the four fundamental physical forces is a synergy, in that the new object is fundamentally different from its constituent parts.

Once the after-effects of the Big Bang settled down, the Universe effectively settled into a steady state. Most of the Universe exists as either a steady state or as a constantly repeating set of processes, the most important of which are stellar nucleosynthesis and stellar evolution.

Biological Evolution

Biological evolution was the next level of evolutionary processes. Biological evolution is based upon combining a sub-set of elements and molecules to form a lifeform that is capable of surviving and reproducing. The key elements are Carbon, Hydrogen, Oxygen, Nitrogen plus a few others. Chemists know these objects as organic elements.

Biological evolution is fundamentally different from Physical/Chemical evolution in that biological organisms can reproduce. Physical/Chemical objects can survive based upon the interaction of the four fundamental physical forces, but they cannot reproduce. The ability to reproduce and change slightly with each gener-

KEY INSIGHTS

Biological evolution rapidly accelerates the complexity and rate of change on planet Earth.

ation makes life far more complex than the relatively simple atoms, molecules and compounds that make up the rest of the Universe.

Biological evolution is a process of change that is driven by three factors:

1. Variation: every individual varies from other members of the same species.
2. Genetic heritability: many of the variations between individuals are inherited from the biological parents.
3. Differential reproductive success: individuals differ in their ability to survive and reproduce.

Variation is injected into the biological world via three methods:

- Meiosis (the process of cell division that creates four genetically-distinct daughter cells).
- Genetic mutations ("mistakes" in the DNA copying process).
- Sexual reproduction (where half of the chromosomes of the mother are combined with half of the chromosomes of the father).

When these factors are combined, biological evolution is the inevitable result. This is just as true for humans as it is for plants and other animals. While simple in concept, biological evolution has resulted in amazingly complex adaptations. These adaptations are strikingly similar to the technological innovations that enable more complex societies to evolve.

In *Principles of Social Evolution* Andrew Bourke lists the major transitions of biological evolution including:

1. Organic molecules combining to form replicating nucleic acid molecules.
2. Replicating nucleic acid molecules combining to form chromosomes.
3. Very simple organisms combining to form single-celled organisms.
4. Single-celled organisms combining to form multi-celled organisms.
5. Multi-celled organisms combining to form social organizations (i.e. flocks of birds, herds of mammals and schools of fish).

Note that each transition increases the size, complexity and diversity of the resulting biological organism. As we will see, this is remarkably similar to the way that important technological innovations increase the size, complexity and diversity of human societies.

Biological evolution rapidly accelerates the complexity and rate of change on planet Earth. Biological entities acquire energy and other nutrients from the environment and harness biochemical processes to create a stable internal system that enables them to survive and reproduce. Because there are finite amounts of energy and nutrients in nature, organisms must compete with each other to survive and reproduce.

The combination of finite resources, variation and differential reproductive success makes the biological world far more dynamic and complex than the Physical/Chemical world. Each generation is better adapted to its environment than the last. And each species within an eco-system interacts with the others. The result has been a constant increase in the number of species and a general movement towards increasing complexity.

Cultural Evolution

After billions of years of Biological evolution, modern humans evolved.

Humans created a new form of evolution: Cultural evolution. Cultural evolution is the change of human societies over time.

The concept of Cultural evolution is one of the hottest topics of the last few decades. Insightful thinkers such as Peter Richerson, Robert Boyd, Joseph Henrich and Alex Mesoudi have given us valuable insights about the concept.

Cultural evolution encompasses all elements of human societies including religion, ethnicity, language and culture. I will mainly focus on innovations to technology, skills or organizations, as they are the most closely related to material progress.

Cultural evolution builds on the previous forms of evolution by adding in conscious design, trial-and-error experimentation and copying. It is no longer necessary to wait for accidental combinations to take place to form a higher synergy.

With Cultural evolution human beings can deliberately create those combinations, assess the usefulness of the results and iterate on those results. Just as importantly, human beings can see the creations of other people, assess their usefulness, copy them and modify them for different uses. Whereas biological evolution is based on survival and reproduction, the "survival" of a technology, skill or organization is based upon its usefulness to humans.

Cultural evolution has many advantages over biological evolution in its ability to adapt to a highly complex and changing environment. In biological evolution, new variations are created randomly. In cultural evolution, humans can design a new variation to solve a problem.

Conscious design enables humans to consider many possible solutions and logically determine which variations have the greatest chance of success. In particular, potential innovators can ignore variations that seem unlikely to succeed.

In biological evolution, new variations are created only once per generation. For simple creatures with short lifespans measured in hours or days, this is not much of a barrier. For complex creatures with long lifespans measuring in decades, however, this is a serious constraint on the rate of innovation.

In cultural evolution, there are no limits on the number of innovations that can occur within one lifetime. Cultural evolution enables a successful innovation to spread rapidly through a society within one generation. This makes it far more likely that one person with the innovation will survive to pass it on to the next generation.

In cultural evolution, humans can share ideas and designs during the experimentation process. This increases the number of variations as well as increasing the success rate. So there is a greater chance that the new technology will be useful.

> **KEY INSIGHTS**
>
> Cultural evolution by humans adds in conscious design, trial-and-error experimentation, teaching and copying. It is no longer necessary to wait for accidental combinations to take place. We can create them.

In biological evolution, an adaptation will only be propagated through a species if it confers an advantage when it comes to surviving and reproducing. In cultural evolution, humans can innovate solutions that do not directly impact their rate of survival and reproduction. They may, for example, create innovations that make their lives healthier and more enjoyable.

In biological evolution, genes cannot choose to share themselves with others. In cultural evolution, humans often share skills with other humans by consciously making it easier for others to copy them. We call this teaching.

Teaching consists of a large number of complex tasks:

1. Explaining important concepts related to mastering a skill
2. Breaking the skill into manageable chunks
3. Demonstrating the proper actions
4. Carefully watching their student attempt to copy those actions
5. Noticing incorrect actions taken by the student
6. Explaining to the student how they can improve
7. Emotionally reassuring the students after they fail, so the student will keep trying.

In biological evolution, a person can only acquire new information from the genes of their biological mother and father. In cultural evolution, a person can acquire new information from a wide variety of people, including:

- Extended family (grandparents, aunts, uncles and cousins)
- Friends
- Neighbors
- Teachers
- Co-workers
- Strangers who use information storage technologies, such as books or the internet.

The wide variety of potential people to copy gives humans far more examples than would be possible if they were restricted to biological evolution.

In biological evolution, people cannot consciously choose whose genes to copy. The genes of children can only copy the genes of their biological parents. In cultural evolution, people can consciously choose whom to copy. This greatly increases the chances that they are copying a useful innovation.

Common strategies include copying:

- The most successful person in a community (for example, a successful merchant, athlete or hunter)
- The most prestigious person in a community (for example, the local lord or celebrity)
- The person most similar to themselves (for example, someone of the same occupation, gender or ethnicity)
- The majority of people in their local environment (hence the classic phrase "When in Rome, do as the Romans do")

Each strategy for copying is likely to lead to a different solution, leading to greater variation within a society. This increases the likelihood that the result will be useful.

Cultural evolution also enables an individual to copy multiple people, each with unique varieties of skills, technologies and behaviors. As I mentioned earlier, studies have shown that a person who copies many different people is more likely to get positive results than an individual who copies only one person.

An individual who copies five different people instead of one can pick specific elements that appear to work for each person, then recombine them with other elements from other sources. In this way, copying leads to new variations with a higher chance of success. Once again we see how copying leads to innovation.

In biological evolution, all information related to survival and reproduction is based upon the local environment. It is possible to be highly adapted to a specific microenvironment, while also being totally unable to survive in another environment.

In cultural evolution, humans in one society can see solutions that work substantially better for another society in a different environment. This is particularly true once transportation and communications technologies evolved enough to enable humans of very different geographical locations to interact regularly with each other.

Biological evolution operates almost exclusively on the individual. While animals can combine into social organizations, these social organizations are usually very simple and composed of other animals that are almost identical to each other. In cultural evolution, humans can form complex social organizations composed of people with widely varying skills. This enables far more complex solutions to evolve.

Biological evolution requires a certain amount of stability in the environment so that optimal solutions can be found. Biological evolution is good at finding optimal solutions within environments that either do not change or change at a slow rate in a consistent direction. Very rapid changes in the local environment can lead to wide-scale extinctions.

Cultural evolution enables humans to adapt to rapid changes in the environment within one generation as well as rapidly swinging directions of change. This makes cultural evolution far more flexible than biological evolution.

Despite all its advantages, it is important to keep in mind that cultural evolution also has some important disadvantages. Because human beings copy those around them, ideas and practices can evolve that are highly detrimental to humans, either as individuals or as societies. We will come back to this phenomenon later.

Conclusion

Progress is the outcome of an evolutionary process that goes back billions of years. This process started with the Big Bang, which allowed Physical/Chemical evolution to start. Eventually, organic molecules combined to form replicating molecules. This triggered Biological evolution. The world became more complex and prone to rapid change.

Then humans developed Cultural evolution, which enabled humans to channel energy into creating new technologies, skills and organizations to solve problems. Humans learned how to cooperate in organizations, and they learned how to compete against other groups. Most of all, they learned to copy the technologies, skills and organizations innovated by others. This enabled each generation to build upon the learning of all the previous generations.

But something was missing. Despite all of this change and all of this innovation, the material standard of living for the masses did not improve. Some key ingredient was missing from the recipe. Or more correctly five key ingredients were missing.

Humanity had not yet found the Five Keys to Progress.

LIFE BEFORE
PROGRESS

We have seen that human societies have gradually overcome the constraints to progress by innovating technologies, skills and organizations for turning the energy of the sun into useful technologies. But how did this happen? How did human societies evolve from small bands roaming the African savanna into complex societies capable of harnessing enormous amounts of energy and complex technologies?

To explain how humans escaped the traps created by constraints to innovation and progress, we must briefly overview human history. In the distant past, change, let alone progress, was not part of anyone's experience. A person's daily life was pretty much the same as their parents, their grandparents and their great grandparents.

The most important changes within a person's lifetime were sudden and negative: droughts, crop failures, famines, wars and epidemics. Once the effects of these deadly events wore off, life returned pretty much to the way it had been… at least for those who were lucky enough to survive.

Traditional Chinese concepts of history understand history as a cycle. Change occurs, sometimes even something resembling progress, but then the cycle reasserts itself to reset things back to the way they were. I think this view adequately represents most of the world up until the 20th Century.

Society Types

Human societies have not been the same across the globe and across history. Mainly because of geographical variation, very different types of societies evolved in different parts of the globe.

While they each had important cultural differences, it is most convenient to categorize societies into what anthropologists and sociologists call "society types." The concept of society type enables us to overview human history without going into all the details that were only relevant in the short term.

A society type is a category for how a society organizes itself to transform energy and other natural resources into food and other useful technologies. This is most easily measured by how people acquire a majority of their calories. If, as they say, "you are what you eat", then the society type concept emphasizes that "we are how we acquire what we eat."

While they did not use the term, the concept of Society Types first became popular during the Enlightenment in the 18th Century. Early thinkers on the subject widely believed that human societies naturally progressed from simple to complex. This created a very linear view of human history. This way of thinking is sometimes called "unilineal evolution of societies."

For example, most Enlightenment thinkers believed that all societies started as Hunter-Gatherers, then at some point transitioned into Herding societies, and then at a later date, they transitioned into Agricultural societies. Adam Smith and some thinkers within the Scottish Enlightenment added on a fourth type of society, Commercial societies, to explain the transition going on at that time. Adam Smith thought that all societies would transition through all the other society types until they inevitably transitioned into modern Commercial societies.

Later in 19th Century, thinkers such as Henry Home, Edward

Burnett Taylor, Lewis Henry Morgan developed more complex versions of the theory. In the 20th Century, Julian Seward and Gerhard Lenski further developed the concept. At one time, Society Type was a dominant paradigm in anthropology and sociology, but it has lost favor within those fields.

> **KEY INSIGHTS**
>
> The concept of society type enables us to overview human history without going into all the details that were only relevant in the short term.

As it became clear that there were other types of societies that do not easily fit into those categories and that many societies failed to transition to "higher" states, the concept fell out of favor. Just as important, many early interpretations during the Enlightenment and 19th Century included strong value judgments with more complex societies being labeled as "civilized", while simpler societies were labeled as "savages" or "barbarians." In addition, global histories that spanned millennia fell out of favor. All of the above meant that the concept of Society Type is not widely used today.

I believe this is an overreaction to the overly simplistic application of early versions of the concept. If we apply an evolutionary perspective, we can see that just as animals adapt to their natural environment, humans adapt to their social environment (i.e. Society Type), which in turn is an adaptation to their natural environment.

None of these adaptations are inherently superior to the others. Specific adaptations promote survival and reproduction in a specific environment. Those same adaptations may not work particularly well in a different environment.

Virtually all energy in human societies ultimately derives from the sun. Societies use different technologies, skills and social organizations to transform solar energy into food that is fit for human consumption. In this way, large groups of people can survive and reproduce. How they do so, however, varies greatly between societies.

When viewed from the highest level, there are only a small number of means through which a society can acquire enough calories to survive and reproduce. The following is a list of society types and how they acquire the majority of their calories:

- Hunter-Gatherer societies: Hunting wild animals and gathering wild plants.
- Fishing societies: Fishing, hunting sea mammals and gathering shellfish.
- Horticultural societies: Farming domesticated plants (using hand tools) and sometimes also raising domesticated animals.
- Agrarian societies: Farming domesticated plants (using animal-driven iron plows) and raising domesticated animals.
- Herding societies: Herding domesticated animals on the wild range.
- Commercial societies: Selling a product or skill, so one can buy food from the market (but with limited use of fossil fuels).
- Industrial societies: Selling a product or skill, so one can buy food from the market and widespread use of fossil fuels.

It All Comes Down to Food

But why is food consumption so important that we can categorize entire societies by it? The reason is that until very recently, by far the most time-consuming task in people's daily lives was in acquiring food. Other than sleep, the bulk of people's time has been allotted to producing, preserving, storing, distributing, preparing and eating food. And much of the time left over has been devoted to building and repairing tools necessary for those tasks.

Because so much time and energy was needed to acquire food, entire societies were forced to organize themselves around the optimum means of doing so within their local environment. Because of environmental limitations, the limited number of skills that any one person can acquire, and the need for people to cooperate within social organizations, societies must specialize in one specific means of acquiring food.

For example, a group of people in one geographical area cannot simultaneously specialize in both fishing and farming. In any one geographical area, one of these means of acquiring food generates more calories of food per unit of work than others, so most people will naturally focus their efforts on that type of food.

This does not mean that an individual human can only focus on producing one type of food. A farmer, for example, may go fishing in the local stream at a time when farm work is not needed. That individual may even specialize in fishing and provide extra protein for the entire village. But the overall consumption of the village and surrounding region would be largely from domesticated plants and animals.

A very large polity may have different regions specializing in different means of subsistence, but each region must specialize in the means that is most effective at generating calories of food per unit of work. The level of technology, available energy and natural resources determines which means of subsistence is most effective.

> **KEY INSIGHTS**
>
> A society type is a category for how a society organizes itself to transform energy and other natural resources into food and other useful technologies. This is most easily measured by how people acquire a majority of their calories.

The vast majority of societies throughout history fit comfortably into one of the types listed above (although polities often span multiple society types). The exceptional societies that cannot be fit into a type have had relatively little impact on history. Societies do not persist on the edges of a type for long (i.e. 55 percent of calories from one type of food and 45 percent of calories from another).

Again, each society must focus its efforts on the most productive means of acquiring food that the environment can provide. To do otherwise threatens the ability of the group to survive and reproduce.

The Usefulness of the Concept of Society Type

While many historians and anthropologists would undoubtedly bridle at such a gross over-simplification of each society, I believe that Society Type is a useful concept for the following reasons.

Almost all bodies of human knowledge use classifications to promote our understanding. Geologists group clusters of years into eons, eras and periods; Chemists group types of atoms into elements on the

Periodic Table; Biologists group individual plants and animals into domains, kingdoms, phylum, classes, orders, families, genus and species.

All of these classifications simplify by categorizing individual items into groups based upon relevant characteristics. All are open to the charge of over-simplification, but all these categorizations have stood the test of time because they are useful. They are useful because they focus on differences that matter while ignoring less important differences.

The concept of Society Type is useful because with one simple label we can communicate a large number of important characteristics about any given society in history. We will cover these characteristics in much greater detail later, but for now, it is sufficient to say that each type has common levels of population size, population density, political structure, economic structure and rates of innovation. And these characteristics differ from societies of other types.

The concept of Society Type enables us to understand why some groups of people evolved culturally at vastly different rates than other groups of people. This is because the society type is to humans what the natural environment is to animals: the critical environmental factor that drives evolutionary change. Among animals, that evolutionary change is entirely genetic. Among humans, that evolutionary change is genetic, cultural and technological.

Humans adapt to their society type through changes to their genes, culture, technology, skills and social organizations. Any factors that fundamentally conflict with survival and prosperity in their society type will be under substantial pressure to change.

With adaptation via innovation and diffusion, each generation is slightly better adapted for surviving, reproducing and prospering in that society type than the previous generation. Within a single lifetime, the changes can seem trivial, but when viewed from the perspective of centuries, these changes can be dramatic.

The concept of Society Type enables us to understand why some nations today are much wealthier than others. Hunter-Gatherer societies are uniformly poor by the standards of any other society. Horticultural, Agrarian, Commercial and Industrial societies were step-by-step wealthier than the previous society. And the gulfs between societies of different types were usually vast in size.

The concept of Society Type also enables us to understand why different ethnicities within the same country are much wealthier than others. People whose ancestors were recently from Hunter-Gatherer societies have had a very difficult time prospering in other society types. Immigrants who descend from peoples that lived in Horticultural societies have also struggled to adapt to living in more complex societies, despite the greater opportunities. People from Commercial and Industrial societies have prospered wherever they have immigrated.

The concept of Society Type enables us to understand the daily lives of most people within a society. We have already seen that until the Industrial era, the overwhelming majority of people in all societies did the bulk of their labor acquiring food. And the processes that they developed to grow, distribute and store food played a powerful role in structuring the entire society. For this reason, food production should define how we categorize societies.

KEY INSIGHTS

The concept of society type is useful because with one simple label we can communicate a large number of important characteristics about any given society in history: population size, population density, political structure, economic structure and rates of innovation.

The society type is to humans what the natural environment is to animals: the critical environmental factor that drives evolutionary change. Among animals, that evolutionary change is entirely genetic. Among humans, that evolutionary change is genetic, cultural and technological.

The food production system largely determines the amount and type of energy that a society can consume. This energy can then be used to innovate new technologies, skills and social organizations that solve key problems.

With relatively inefficient food production, very little energy can be devoted to innovation. Virtually all of it is consumed by survival and reproduction. But as the food production system becomes more efficient, then greater amounts of energy can be devoted towards the innovation of new technologies, skills and social organizations.

Overview of Human History

This graphic is a very abstract overview of human history. It displays the link between geography, technology and society type. As we will see later, the society type plays a key role in determining the ability of a specific society to innovate new technologies, skills and social organizations as well as copy the innovations of other societies.

As we survey human history and geography in the next few chapters, I encourage you to repeatedly refer back to this graphic.

Across the bottom of the graphic (x-axis) are biomes, each with color running vertically. I will explain biomes in more detail in later chapters, but for now it is enough to think of them as the dominant vegetation in any one geographical area.

On the vertical axis (y-axis) is the time period with the most recent at the top and ancient history at the bottom. By starting at the bottom of the graphic in the "Apes & Early Hominids" box and then working our way upwards following the black arrows, we can get a high-level overview of human history.

Each society type is displayed as a white horizontal block demarking which biomes that it can evolve within. In addition, the diagram includes a short list of key enabling technologies (in boxes with light gray background) that enable societies to transition from one society type to a more complex society type. The thick black lines connecting the enabling technologies with the society types also show in which biomes that first transition occurred in history.

Looking at this diagram, we can make a few observations:

- Humans have invented only a handful of society types. There are only so many ways for large groups of people to produce enough food to survive and reproduce in a specific natural environment. This limits the number of society types that can exist.
- Each society type can only inhabit a portion of all the biomes on

earth (although Hunter-Gatherer societies come close to being universal). The vegetation that grows naturally in a biome makes many society types difficult or impossible to evolve because the necessary plants and animals do not exist in that biome.

- Even if society types are somehow able to evolve in the "wrong" biome, those societies will compete with other societies that focus their food production efforts on the more cost-effective food sources. This will most probably cause their ultimate failure to survive.

- The invention of key technologies enabled societies to transition from one society type to another society type. These technologies were not invented at the same time so the transition was not instantaneous. In comparison with how long the previous society type lasted, however, the transition was rapid. This transition is akin to the rapid phase transition that occurs when water has been slowly heated and then suddenly turns into steam.

- The transition between society types can only occur in specific "gateway biomes" — biomes that can support both the pre-transition type and the post-transition type.

- Societies that evolved in other biomes are effectively trapped in their society type. The biomes that can support the most technologically complex type in any one era differ, so many societies reach evolutionary dead ends. We will explore this concept in more detail later.

Transforming Energy into Complex Societies

At a very high level, the process through which humans transform energy into complex societies is the following:

1. The Sun creates solar energy by fusing hydrogen into helium. Sunlight injects a portion of that energy onto planet Earth.
2. Plants gather that solar energy and transform it into sugar (via photosynthesis).
3. Herbivores eat plants and transform that sugar into energy to power their bodies (via metabolism).

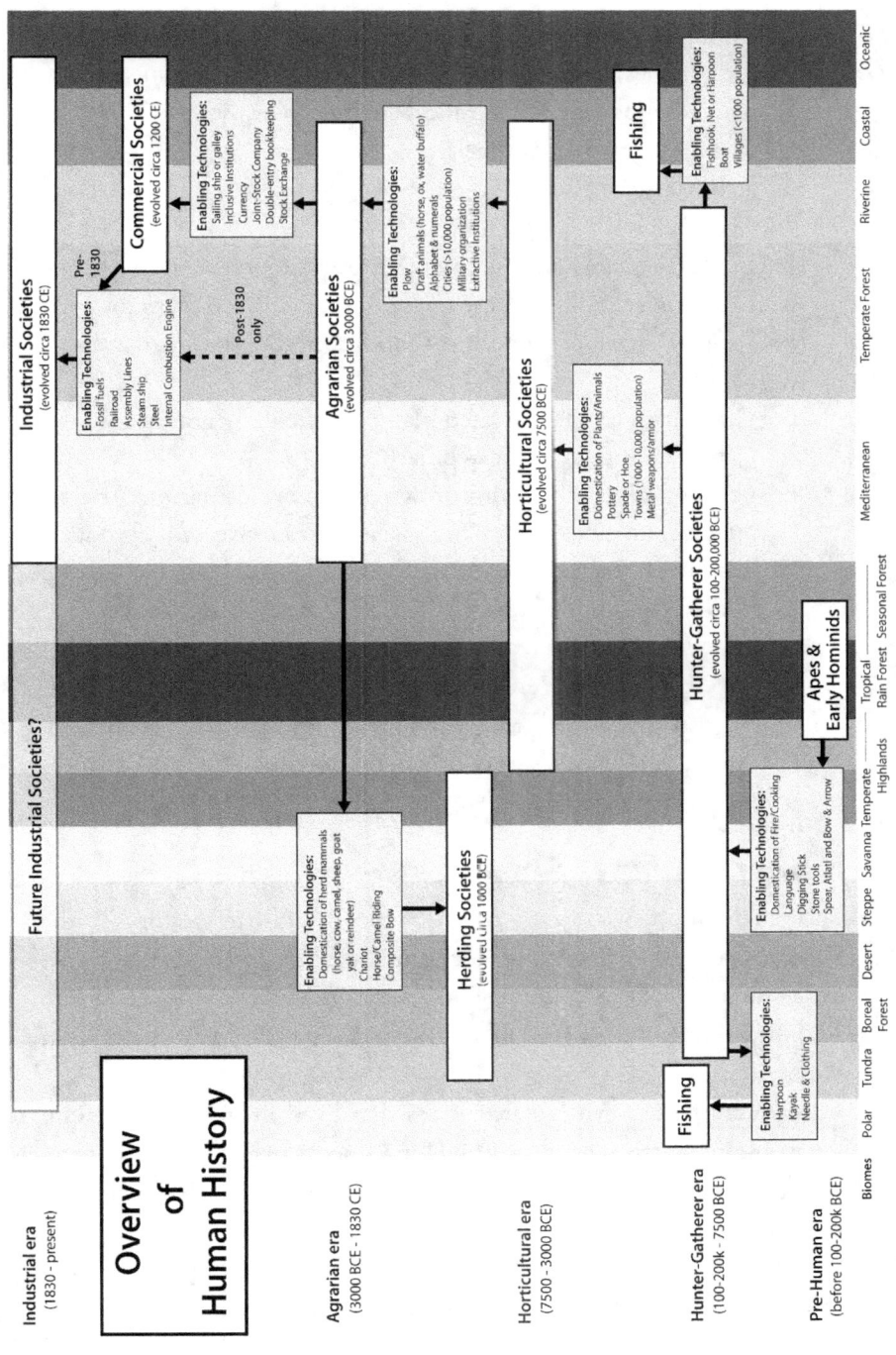

Overview
of
Human History

Industrial era
(1830 - present)

Agrarian era
(3000 BCE - 1830 CE)

Horticultural era
(7500 - 3000 BCE)

Hunter-Gatherer era
(100-200k - 7500 BCE)

Pre-Human era
(before 100-200k BCE)

Industrial Societies
(evolved circa 1830 CE)

Future Industrial Societies?

Commercial Societies
(evolved circa 1200 CE)

Enabling Technologies:
Sailing ship or galley
Inclusive Institutions
Currency
Joint-Stock Company
Double-entry bookkeeping
Stock Exchange

Enabling Technologies:
Fossil fuels
Railroad
Assembly Lines
Steam ship
Steel
Internal Combustion Engine

Pre-
1830

Post-1830
only

Agrarian Societies
(evolved circa 3000 BCE)

Enabling Technologies:
Plow
Draft animals (horse, ox, water buffalo)
Alphabet & numerals
Cities (>10,000 population)
Military organization
Extractive Institutions

Fishing

Enabling Technologies:
Fishhook, Net or Harpoon
Boat
Villages (<1000 population)

Horticultural Societies
(evolved circa 7500 BCE)

Enabling Technologies:
Domestication of Plants/Animals
Pottery
Spade or Hoe
Towns (1000-10,000 population)
Metal weapons/armor

Hunter-Gatherer Societies
(evolved circa 100-200,000 BCE)

Apes &
Early Hominids

Enabling Technologies:
Domestication of herd mammals
(horse, cow, camel, sheep, goat
yak or reindeer)
Chariot
Horse/Camel Riding
Composite Bow

Herding Societies
(evolved circa 1000 BCE)

Enabling Technologies:
Domestication of Fire/Cooking
Language
Digging Stick
Stone tools
Spear, Atlatl and Bow & Arrow

Fishing

Enabling Technologies:
Harpoon
Kayak
Needle & Clothing

Biomes

Polar · Tundra · Boreal Forest · Desert · Steppe · Savanna · Temperate Highlands · Tropical Rain Forest · Seasonal Forest · Mediterranean · Temperate Forest · Riverine · Coastal · Oceanic

4. (In Hunter-Gatherer and Fishing Societies) Humans innovate technologies, skills and organizations that enable them to acquire food from their local environment by:
    a. Hunting wild animals
    b. Gathering wild plants
    c. Fishing
5. (In Horticultural and Agrarian Societies) Humans innovate technologies, skills and organizations that enable them to acquire far greater amounts of food from their local environment by:
    a. Farming domesticated plants, particularly grains.
    b. Herding domesticated herbivores.
6. Increased food production means that a typical farming family can generate more food than they need to survive and reproduce.
7. This increased food surplus is channeled into:
    a. Having more babies, which increases the population. This trend increases the rate of innovation.
    b. Establishing centralized institutions that extract the food surplus from farmers and giving it to political, economic and religious elites. This trend decreases the rate of innovation.

Hunter-Gatherer Societies

The single most important society type in human history is the Hunter-Gatherer society type. For virtually all of our history as a species, humans lived as Hunter-Gatherers. We evolved biologically to adapt ourselves to the needs of surviving and reproducing within that society type.

Human beings evolved from apes that lived in the Tropical Rain Forest biome. Grasses as species had long since evolved, but the climate was too warm and wet for them to become the dominant vegetation in any one area.

Like most large mammals, apes lived in very low population densities (at least in comparison to modern humans). Our ape ancestors possessed very little technology beyond sticks.

As the climate cooled and dried 4-6 million years ago, the grasses gradually grew more common in East and South Africa and eventually they become common enough to dominate entire regions. This marks the birth of the Savanna (or Tropical Grassland) biome.

The Savanna biome in East and South Africa advanced at the expense of the Tropical Rain Forest biome, and early hominids evolved to adapt to the thinning of trees. At some point, key enabling technologies such as the domestication of fire, language, stone tools and weapons evolved in tandem with early hominids.

This tool kit of technologies enabled the evolution of modern humans as a species, as well as the Hunter-Gatherer society type around 100,000-200,000 years ago. Key technologies enabled Hunter-Gatherer societies to do the following:

- Manipulate their environment (using stone tools)
- Dig for underground roots, rhizomes, tubers and bulbs (using digging sticks)
- Communicate complex ideas with each other (using language)
- Safely kill large, dangerous mammals from a distance (using spear, atlatl, and the bow and arrow)
- Extract the maximum amount of nutrition from those mammals with as little work as possible (using fire and cooking)

Hunter-Gatherers Migrate

Hunter-Gatherers originated in the African Savanna, but gradually spread to other eco-regions. As these Hunter-Gatherer societies migrated and encountered new natural environments, they were forced to innovate new technologies to survive. This created new varieties of Hunter-Gatherer societies, each based upon their dominant means of subsistence (i.e. their main diet and how they produced that food).

Hunter-Gatherer societies had access to very limited amounts of energy, so their populations stayed small and bands were widely scattered. The combination of a small base of technology, small population size and low population density meant that the rate of innovation during this period was extremely slow. Despite this, Hunter-Gatherer societies found a way to expand into almost all of the biomes on planet Earth except the open oceans.

As the first Hunter-Gatherer societies evolved, they presumably increased in population. As the local population increased up to the carrying capacity of their environment given their current technology, sub-bands splintered off and moved into virgin territory. Most likely, the younger members of the groups led these sub-bands once they grew into adulthood. Gradually, generation after generation, each band would fill in every geographical niche within the biome.

> **KEY INSIGHTS**
>
> Hunter-Gatherer societies survive by hunting wild animals and gathering wild plants. Their key technologies are stone tools, digging sticks, language, bow-and-arrow, fire and cooking.

This probably continued until Hunter-Gatherer bands reached a major geographical barrier such as an ocean, inland sea, major mountain range or a different biome. When a band reached a different biome, its expansion would have slowed down as they needed time to adapt their subsistence (food acquiring) technologies to the new environment. For example, many of the survival skills and technologies that were highly useful in the Savanna were not particularly useful within Tropical Forests. New skills and technologies had to be innovated.

Most likely these bands remained in the transition zones where the vegetation was a mixture of the old and new biomes. With low population density, Hunter-Gatherers would naturally avoid these less productive areas.

But with increasing population density, some bands would be forced to push further into the transition zone so they could survive. The bands living in these transition zones would be under immense pressure to innovate new technologies and skills that would enable them to exploit all of the resources of the transition zone, not just the ones that they were used to.

At some point in time, the adaptations became sufficient for a full transition to the new biome. When this occurred, a rapid expansion of human populations within that biome probably took place. This process continued for tens of thousands of years until all eco-regions that could support Hunter-Gatherer societies were inhabited.

In this history of Hunter-Gatherer societies, one can see many of the trends that enable progress today. Hunter-Gatherers cooperated to solve local problems. In doing this, they innovated new technologies and skills to make, use and repair those technologies. As the technologies became more complex, Hunter-Gatherers were forced to cooperate more closely together. When Hunter-Gatherer bands competed with each other for territory and food resources, then cooperation within the group became even more important.

Hunter-Gatherers Today

History has not been kind to Hunter-Gatherer societies. While they dominated the planet for at least 100,000 years, they have been in full-scale retreat over the most recent 10,000 years. Today they only live in small isolated regions that are of little interest to other societies. Their small population size and relatively simple toolset has left them highly vulnerable to conquest from more complex societies. While this fate is sad, it is difficult to see how it could have ended up any other way.

When viewed over the span of millions of years, we can see that apes gradually evolved genetically, culturally and technologically into Hunter-Gatherer societies. In any one year or even century, the changes would have been so minute that they were undetectable, but over time great changes took place.

So was there any true progress during the hundreds of thousands of years that humans lived in Hunter-Gatherer societies? While new technologies and skills evolved, it is difficult to see the standard of living of the people improving noticeably within the course of one generation.

The fundamental problem was that humans started from a very small technology base, so the possible number of new combinations that would result in useful technologies was very small. In addition, humans were so spread out in the search for food that they could not come together in the dense populations that are necessary to spur innovation. Finally, because Hunter-Gatherer societies were highly mobile and they needed to physically carry all their technology with them, there was a hard ceiling on the amount of technology that could be acquired.

Hunter-Gatherer societies were problem-solving engines, but of a very inefficient kind compared to what came later.

Recent Genetic Evolution of Humans

This lack of progress does not mean that our history as Hunter-Gatherers is unimportant. There is, in fact, an entire discipline of social science based upon the assumption that it mattered a great deal. Evolutionary psychologists believe that human beings evolved genetically to survive and reproduce in Hunter-Gatherer societies, and much of what we can now call human nature is due to those adaptations.

Evolutionary psychologists believe that both our physiology and our psychology are adaptations to our ancestral environments. Because humans lived in Hunter-Gatherer societies for virtually all of our history, it stands to reason that this social environment had a powerful effect on our bodies and brains. Since agriculture is only 10,000 years old and industry is only a few centuries old, evolutionary psychologists believe that more recent environments have had far less impact on our bodies and brains.

Overall, I agree with the evolutionary psychologists that our long history as Hunter-Gatherers has left a profound impact on humans. Where I differ is that I believe that more recent society types have also had a powerful impact on our bodies and brains.

Until recently, most people assumed that human genetic evolution stopped about 10,000 years ago. In their book *The 10,000 Year Explosion*, Gregory Cochran and Henry Harpending argue that more recent genetic research shows that human genetic evolution over the last 10,000 years has accelerated 100 times faster than in the previous 6 million years.

This should not be surprising. There is strong evidence that evolution among animals tends to accelerate as the population increases and their environment changes. Since both of those factors have increased dramatically over the last 10,000 years for humans, one would have to assume that this would lead to genetic change among humans as well. That is exactly what genetic researchers find.

> **KEY INSIGHTS**
>
> Human genetic change is accelerating.

In addition, human groups are evolving in different directions. Among the most important recent genetic adaptations that have so far been discovered are:

- Changes in the shape of the inner ear, which is necessary for processing language
- Smaller jaw and larger cranium to make room for a bigger brain
- Regulation of blood sugar to protect against diabetes
- Ability to digest cow's milk
- Ability to metabolize alcohol
- Lighter skin to aid Vitamin D production
- Immunity to new infectious diseases, such as measles, smallpox and malaria.

All of the above genetic adaptations have taken place within the last 10,000 years and most have occurred in some human groups but not others. Since genetic research is still in its infancy, it seems likely that scientists will continue to find evidence of recent genetic change. The traditional assumption that the genetics of all current human populations are unchanging and identical is incorrect.

The Legacy of Hunter-Gatherer Societies

Compared to virtually all other peoples, societies that still live as Hunter-Gatherers are impoverished. That is hardly surprising given their very low levels of technology. But even when recent ancestors abandoned the Hunter-Gatherer lifestyle and live in modern societies, they have had trouble succeeding.

While there is very little data on the subject, I can think of no ethnic group whose ancestors lived as Hunter-Gatherers in the year 1500 that are above average in socio-economic status in any nation of the world. To the best of my knowledge, they are all very low in their socio-economic status. Adopting technology and culture alone cannot raise the ancestors of recent Hunter-Gatherers up to the rest of society.

I believe that people living in Hunter-Gatherer societies have evolved genetically, culturally and technologically to survive and reproduce within those societies. Since the means to prosper in those societies is fundamentally different from the means to prosper in modern Industrial societies, they have trouble succeeding in those societies. It has nothing to do with race or inferiority; it has to do with being better adapted to a different environment. And that environment includes both the natural environment and the human environment (i.e. the society type).

To prove that this difference has nothing to do with superiority, think of yourself or some other person living in a modern Industrial society suddenly being transported back into Hunter-Gatherer times. If we lived alone, we would likely die within days or weeks.

If we were lucky enough to be transported into a sympathetic Hunter-Gatherer band, we would at first be as helpless as children. Perhaps after a few months of patient teaching by experienced Hunter-Gatherers, we could begin to contribute to the band in a meaningful way. After a year of learning, we might even be as productive as the other members of the band.

But all the knowledge, skills and beliefs that help us survive and prosper in modern Industrial societies would be almost useless. We would be forced to learn everything from scratch and would only have a chance to do so if others who were already adapted to that environment were willing and able to teach us.

Fishing Societies

As they migrated across the planet, Hunter-Gatherer bands encountered large bodies of water such as coastlines, rivers and lakes. In these areas, humans discovered opportunities to fish, gather shellfish and hunt marine mammals.

Most likely, it took them some time to figure out how to most effectively exploit those new resources, but humans had a strong incentive to do so. Fish and marine mammals are much denser sources of energy than plants or land animals. And many of them can be harvested with relatively little physical effort.

Key technologies enabled Fishing societies to do the following:

- Catch wild fish (using hooks, nets and weirs)
- Transport themselves and their catch over large bodies of water (using kayaks and boats)
- Hunt wild sea mammals (using harpoons)

Fishing societies had much larger populations that were more densely populated than Hunter-Gatherer societies. Those with stable food sources had a strong incentive to stay there, so the settlements became the first villages. While it has long been thought that the sedentary lifestyle started with agriculture, more and more archeological digs are turning up evidence of fishing villages that preceded agriculture by millennia. The larger population sizes, greater population density and sedentary lifestyle probably accelerated technological innovation in Fishing societies.

As the first Fishing societies evolved, they presumably expanded rapidly along the original body of water. Because Fishing societies have larger populations, it was probably fairly easy for them to displace Hunter-Gatherer societies along the body of water. This probably continued until the entire body of water that possessed adequate fishing resources was filled with separate Fishing societies.

Fishing societies were probably also the first society to evolve the chiefdom, a highly personalized style of leadership. With increased food production and a population that was tied to a specific water source, chiefs could extract surplus food for their own needs and disperse that surplus to their loyal followers. This created the first political inequality.

Particularly in areas where the fish were abundant and highly seasonal, for example, salmon in the Pacific Northwest of North America, year-round food storage became important. Chiefs gained control over these storage locations, giving them greater control over the rest of the population.

So did Fishing societies experience progress? While there might have been a burst of progress when Hunter-Gatherer bands first encountered bountiful fishing spots and learned to innovate technology to capture, store and prepare this new type of food, it is doubtful that it could have lasted for very long.

Because Fishing societies were sedentary and lived in more densely populated villages, there were greater opportunities for innovation. But given the very small technology base, it is difficult to believe that members of Fishing societies would experience noticeable progress within one generation.

If our planet had an unlimited number of fishing spots with an unlimited number of edible fish, Fishing societies might have generated a fair amount of progress for humanity. Unfortunately, regions that can support Fishing societies are relatively rare, generally small in geographic extent and available fish stocks are finite. Fishing became a niche society type rather than a true pathway to progress.

Because Fishing societies were heavily constrained by their need to remain near large bodies of water, they were unable to expand beyond very narrow geographic confines. This meant that Hunter-Gatherer societies and Fishing societies coexisted for tens of thousands of years in separate areas.

Fishing Societies Today

Just as with Hunter-Gatherer societies, Fishing societies have been in relentless retreat for the last 10,000 years. Fortunately for them, Fishing

> **KEY INSIGHTS**
>
> Fishing societies survive by catching wild fish, gathering shellfish and hunting marine mammals. Their key technologies are fish hooks, nets, weirs, boats, kayaks and harpoons.
>
> Fishing societies likely invented the first year-round villages long before agriculture was invented.

societies have been able to maintain some semblance of their traditional ways. Many have gradually transformed into seem-autonomous fishing villages within a larger agricultural society. Most, however, have long since been absorbed into more populous societies without a hint of their prior existence.

Fishing societies have been driven back to the Arctic regions, Pacific Northwest and small coastal, river and lake locations. Many of them have been engulfed by more complex agricultural societies, but because their traditional lifestyle required very little land, they could maintain their ways.

And just as the descendants of Hunter-Gatherer societies experience a much lower standard of living even when they live within modern Industrial societies, so do the descendants of Fishing societies. I can think of no ethnic group descendant from Fishing societies in the year 1500 that is above average in socio-economic status in any nation of the world.

Society Types Are Not Moral Judgments

One the most common arguments against the concept of society types today is that the concept confers a supposed moral superiority of some societies over other societies due to them being more advanced. While many thinkers during the Enlightenment and the 19th Century thought some human societies were morally more advanced than others, there is no reason to do so today.

This moralistic framework is categorically false. Society types are about the survival and reproductive of entire societies. A society is either adapted to its natural environment and the people living within them or it is not. Members of that society may be perfectly adapted to living within that type of society, but when they go to another type of society or another geographical environment, they will literally become "fish out of water" (i.e. in grave danger of not surviving).

Rather than using terms such as "advanced" and "primitive", I use the terms "complex" (i.e. with a great deal of internal differentiation and sub-groups) and "simple" (i.e. with relatively small amounts of internal differentiation). This way, there is no moral judgment conferred, no more than when chemists talk of simple or complex compounds of chemical elements.

As I will explain later, Commercial and Industrial societies have an internal dynamic that creates progress. I believe that this progress leads to such immense material benefits for humanity, that they are morally preferable to other society types.

But even that argument makes no moral judgments between Agrarian, Horticultural, Herding, Fishing and Hunter-Gatherer societies. These different types of societies are just different adaptations to a specific natural geography given the technology that had existed at the time.

Society Types Are Adaptations

In his outstanding book *The Secret of Our Success* Joseph Henrich gives a compelling example of how difficult it is for people living in one society type to transfer their success to another environment. In 1845, two British ships searching for the Northwest Passage were shipwrecked on King William Island in northern Canada. No one knows their full story, but we do know that none of them survived.

The sailors on the ship were members of the richest and most technologically dynamic society at that time. Their ancestors had innovated an incredible number of technologies, learned new skills and cooperated in highly complex organizations.

The British at that time were at the pinnacle of human achievement. But the skills, technology, values and organizations that enabled the British to prosper in London or onboard ships were almost useless to survive on an island in northern Canada. None of the crew of 129 survived the ordeal.

At first impressions, this should not be a surprise. Northern Canada is a very hostile environment. No one could survive such a hostile environment without help from outside.

One might think so, but the local Inuits had survived and prospered in that land for about 800 years. The technologies, skills, social organizations and values of the Inuits were very simple in comparison to what the British had innovated, but that did not matter.

What mattered is that the Inuit technologies, skills, social organizations and values were perfectly adapted to that polar environment. Inuits were neither inferior nor superior to the British. If Inuits had suddenly been transported to Britain, they would have struggled as well. Perhaps they would have been able to survive better than the shipwrecked British crew, but it is doubtful that they would have done as well as a typical British person.

Every Inuit child was taught from birth the skills, social interactions and values needed to survive and reproduce in that polar environment. This made each child the benefactor of centuries of trial-and-error experimentation by their ancestors that led to key technological innovations such as kayaks, snow houses and cold-weather clothing.

Inuit boys were taught by their fathers how to hunt seals, build and repair kayaks and spears, predict the weather and a myriad of other skills. Inuit girls were taught how to preserve and prepare food and tailor cold-weather clothing. Both girls and boys learned how to cooperate in the types of social groups necessary to survive in the polar environment.

Over the generations, Inuits built Fishing societies adapted to survive in environmental conditions that would terrify most other people. This did not make them superior, just better adapted to a specific local environment.

If forced to live in a very different natural or social environment, the Inuits would have had a very difficult time surviving. Even if they somehow were able to survive and thrive, it would be because they were able to rapidly copy members of a society that were already better adapted to the new environment.

Horticultural Societies

As Peter Bellwood and many others argue, one of the greatest innovations in human history was learning how to cultivate grain. Humans can consume a remarkable range of plants. We consider approximately 7000 plant species to be edible.

Despite this diversity, the majority of the calories consumed by humans throughout the last few millennia has been from a very small list of staple crops: rice, wheat, corn and potatoes. If one adds in a few more related cereals like oats, rye, barley and millet, a slightly longer list of foods has supplied over 80% of global food needs.

Except for potatoes, all of these crops are closely related to each other. They are all cereals from the Poaceae family, more commonly known as grass.

It seems highly likely that Hunter-Gatherer societies had always gathered wild

> **KEY INSIGHTS**
>
> Horticultural (or gardening) societies survive by farming domesticated plants using hand tools and sometimes also raising domesticated animals. Their key technologies are the spade, hoe and other hand farm tools along with pottery, granaries and bronze weapons/armor.

grain where it was available. Unfortunately, outside of Temperate Grasslands, most ecosystems lacked this crucial energy source. Around ten thousand years ago in the Hilly Flanks of modern-day Iraq, Syria and Southern Turkey, some societies began to experiment with harvesting wild grain (such as wheat and barley) on a much larger scale.

These grasses packed far more energy per unit of labor and grew across far wider swaths of land than any existing food sources, so many people sought to harvest these wild plants. As the scale of production increased, these societies shifted consumption heavily towards these plants and invented new technologies to cultivate and store these grains.

The people of the Hilly Flanks also played an important role in domesticating cattle, goats, pigs and sheep. These animals were highly important to survival, as they possessed the ability to digest cellulose, the dominant portion of grains. While humans could neither eat nor digest the cellulose that was all around them, they were able to eat the cattle, goats, pigs and sheep that could do so. In this way, humans innovated a means for turning a very widespread natural resource into food.

The people of the Hilly Flanks devoted more and more time to cultivating grasses and domesticated animals, forcing fundamental genetic changes in both the grasses and themselves. Most importantly, these behavioral changes forced people to evolve the ability to digest glucose and build up biological resistance to diseases transferred from domesticated animals. People domesticated the grasses and herd animals, but the grasses and farm animals also domesticated the people.

Eventually, the Horticultural society evolved, in which the bulk of people were engaged in harvesting domesticated grasses, animals and other staple crops. Unlike later Agrarian societies, they lacked animal-driven plows, so all of the work had to be done by simple hand tools. You can think of these people as gardeners, who used methods very similar to what people today use in their backyard gardens.

Key technologies enabled Horticultural societies to do the following:

- Domesticate wild grains, particularly rice, wheat, sorghum and millet (using spade, hoe and other hand tools)
- Domesticate wild herbivores, particularly cows, goats, pigs and sheep.
- Store and transport their grain (using pottery and granaries)
- Wage war against other humans (using bronze weapons and armor)

This suite of new subsistence technologies for growing, storing and transporting domesticated grasses created such a large energy surplus that much of it could be used to support the earliest cities and towns.

Horticultural societies generated far more calories per human than Hunter-Gatherer societies, and they were not restricted to rivers and coastlines as Fishing societies were. This meant that Horticultural societies could expand far beyond their original borders.

Horticultural Societies Migrate

Except in regions along large rivers or where the soil was very rich, Horticultural societies practiced slash-and-burn hand tilling that ruined the local soil productivity within a few generations, so farmers always had an incentive to move to new areas. With simpler technology and much smaller population sizes, neighboring Hunter-Gatherer societies were at a distinct military disadvantage. Migration by farmers in search of fertile soil gradually forced Hunter-Gatherer societies to retreat until they fell back to regions that were less hospitable to slash-and-burn gardening.

Horticultural societies eventually spread to most regions within the Mediterranean, Temperate Forest, Tropical Forest or Tropical Highland biomes. This created a broad belt of Horticultural societies running from East Asia to Europe, as well as a few isolated areas on other continents such as New Guinea, the African Sahel, Mesoamerica, Amazonia and the Andes.

The populations of these Horticultural societies became particularly large and dense along large river valleys, such as the Nile, Tigris/Euphrates, Indus and Yellow Rivers. The rest of the world, however, still lived in Hunter-Gatherer and Fishing societies, because these people lived in environments that could not sustain Horticultural societies.

By the year 1000, most Horticultural societies in Eurasia had already transformed into Agrarian societies. Wherever the environment enabled animal-driven iron plows, purely Horticultural societies had gone extinct.

The bulk of the Horticultural societies in 1500 were in Eastern North America, Central America, South America, Africa, New Guinea and mountainous parts of Southeast Asia. All of these areas lacked the wild ancestors of horses, cows, and water buffalo that were used to pull iron plows.

So the invention of agriculture was one of the greatest innovations of human history, but did it lead to progress that benefitted the masses? It certainly increased their ability to produce additional calories. It also enabled the growth of large cities and expanded the technology base. For the first time, a portion of humanity could focus on solving problems other than producing food.

Unfortunately, the biggest impact of this increased food surplus was to have more babies, which created more mouths to feed. So in the long run, the increased wealth led to a larger population size rather than a wealthier population. In addition, political leaders, religious leaders and warriors probably expropriated the bulk of the food surplus. Even worse, diseases transmitted from domesticated animals were a constant threat.

The Horticultural era resulted in progress for the elites, but it is difficult to find much evidence of progress for the masses. The formation of cities undoubtedly accelerated the rate of innovation and made it much easier for those innovations to diffuse throughout urban populations, but it is clear that the benefits were largely enjoyed by the elites.

The Legacy of Horticultural Societies

The legacy of Horticultural societies still lives with us today. The ancestors of Horticultural societies in the year 1500 (i.e. just before the European conquests) have struggled to prosper. Native Americans in both North and South America, Sub-Saharan Africans, New Guinea and the mountainous parts of Southeast Asia are all either very poor nations or relatively poor minorities within more prosperous nations. To the best of my knowledge, no significant ethnic group descended from Horticultural societies in 1500 is above average in income or social status within their nation.

Agrarian Societies

As the population increased in Horticultural societies, humans were under constant pressure to innovate more efficient means of producing, preserving, storing and distributing food. In a few major river valleys around 3000 BCE, animal-driven scratch plows evolved that increased agricultural productivity far beyond the level possible with simple hand tools.

By substituting more powerful animals (usually horses, oxen or water buffalo) for human muscle power, these societies created far larger food surpluses. At first, these societies were dependent upon irrigation to grow crops, but some farmers learned how to use plows in more arid regions. This key enabling technology gave birth to Agrarian societies in the Middle East and later throughout much of Eurasia.

Agrarian societies dominated recorded history up until the 19th Century. Such societies, including the Babylonians, Sumerians, Hittites, Chinese, Koreans, Japanese, Indians, Greeks, Macedonians, Romans, Byzantines, Persians, Ottomans, Spanish, Portuguese, French, Germans and Russians almost exclusively monopolize our pre-modern history books.

KEY INSIGHTS

Farmers rapidly remove nutrients from the soil, forcing them to migrate to new lands. This pushes Hunter-Gatherers out of their territory.

These Agrarian societies produced large cities, powerful armies, sprawling empires, impressive architecture, beautiful art, compelling religions, and most of all a written record of all of these achievements. Sadly, though, Agrarian societies did relatively little to benefit the vast bulk of their inhabitants.

Key technologies did enable Agrarian societies to do the following:

- Dramatically increase the productivity of their agriculture (using draft animals to pull iron plows)
- Store information and communicate at long distances (using the alphabet, numerals, writing and printing)
- Extract the food surplus from peasants (using centralized institutions)
- Wage war against other societies (using armies and navies equipped with iron weapons and armor and later gunpowder)

While Agrarian societies had a vast number of cultural differences, they had many common characteristics. Most importantly, they produced the majority of their food calories from animal-driven plows. The plows evolved over time. Starting from simple scratch plows that used a wooden post to just scratch the surface of the soil, iron shares were gradually adopted. This gave the plow far greater strength to break tougher soil, expanding the area that could be cropped.

The animals used to pull the plow also varied. East and South Asia generally used water buffalo, while the Mediterranean and Europeans used oxen and later horses. Both are pathetically slow and weak compared to modern tractors, but compared to hand hoeing, they were a huge time-saver. By plowing land faster, peasants could then work larger plots of land, leading to greater grain production and more mouths to feed.

All Agrarian societies relied predominantly on one staple food: The Middle East, Mediterranean and Europe relied mainly on wheat, while East, Southeast and South Asia relied mainly on rice (although the early Chinese societies first relied on millet, before switching to rice). Rice and wheat both produce seeds that are packed full of carbohydrates, as well as some protein and vitamins.

These grains were easily harvested, stored and transported. This made it possible for Agrarian societies to accumulate the huge stores of food that are necessary for building cities and armies. Without rice and wheat (both of which are grasses), none of this would have been possible.

Because of the bounty from these simple grasses, Agrarian societies all had large, dense populations compared to other simpler societies. Productive agriculture meant more people. Unfortunately, population size quickly expanded beyond the level that the local land could support, so there were always hordes of potential settlers eager for new land. Some found those lands by draining swamps, building irrigation to extend rivers into arid lands, or cutting down forests. The desire for land and the food that could be produced on it was never-ending.

Agrarian societies had an extraordinary level of social stratification. Each had a highly developed system of classes, orders or castes. Each group separated people based on how they earned a living. Peasants were invariably at or near the bottom. The king or emperor was invariably at the top of the social order. Beneath him were royal bureaucrats, titled nobility, landowners, religious leaders and military officers. Merchants and artisans were invariably near the bottom, sometimes even below the peasantry, for example in China.

Elites in Agrarian societies were obsessed with status. Each class was denoted by its unique way of dressing, mannerisms, hairstyles, spoken accent, architecture and more. All of this communicated the message that the membership of each class was closed, immortal and not subject to questioning. Many Agrarian societies closely tied these classes to religious beliefs that gave the hierarchy even greater legitimacy.

Agrarian societies were similar to Horticultural societies in that both were heavily reliant on agriculture for their food. Agrarian societies were distinct, however, in having large cities with workers specializing in a wide variety of skills. The combination of large food surpluses and specialization of skills enabled a wide variety of political, economic, military and educational institutions to emerge for the first time.

> **KEY INSIGHTS**
>
> Agrarian societies survive by farming domesticated plants using animal-driven plows and raising domesticated animals. Their key technologies are draft animals, plows, writing, printing, extractive institutions, iron weapons/armor and later firearms.

These institutions concentrated power in the hands of a small elite in huge capital cities: Beijing, Paris, Berlin, Vienna, Moscow, and Baghdad. This elite acquired their wealth by extracting resources through taxation and land rents. The conspicuous consumption of these elites, a means of flaunting their social status, had the side benefit of supporting large numbers of scribes, artisans and traders.

Agrarian societies were not the type of societies that would appeal to many modern readers. The peasants did all the work, while the elites extracted the benefits of that work. The few people in between were entirely dependent upon finding favor from elites to get their little piece of those extracted resources.

Because of their dependence on agricultural production, Agrarian societies were extraordinarily rural. Typically, 97 percent of the population lived in rural areas, hamlets or small villages with a population of under 10,000. The remaining three percent of the population tended to live in one large capital city or one of the regional capitals.

Expansion of Agrarian Societies

One of the most important institutions in Agrarian societies was the military. In previous society types, every able-bodied young man served as a warrior. In Agrarian societies, warriors gradually morphed into a specialized warrior class and finally into professional soldiers and officers.

Armies with bronze, and later iron weapons and armor could conquer large swaths of land that were conducive to animal-driven plows. Except in Tropical regions that could not support plow-based agriculture, a handful of very powerful Agrarian empires absorbed a constellation of smaller Horticultural societies. Only in the Andes, Mesoamerica, New Guinea and Sub-Saharan Africa were Horticultural societies able to survive, aided mainly by their geographical isolation.

By conquering new lands, Agrarian societies enabled settlers who were hungry for land to migrate into these new territories. Sometimes they settled in previously unused land, while sometimes they used violence to push the previous inhabitants out. But wherever they migrated, the settlers brought with them their genes, technologies, skills, values and social organizations.

Unlike less complex societies, Agrarian societies could expand into the periphery of neighboring biomes because they had complex military organizations that could support long supply chains. Even if Agrarian societies could not grow crops in a region, they often coveted other critical natural resources, giving them a strong incentive to conquer and colonize less complex societies.

This meant that Agrarian empires were able to conquer large swaths of land that were not suitable for plow-driven agriculture. By the year 1 Agrarian empires had conquered and settled a huge swath of the Eurasian continent stretching from China, through India, the Middle East, the Mediterranean and parts of western Europe.

How Agrarian Societies Stifled Innovation

Though Agrarian societies had both cities and more efficient food production systems than previous societies, these societies tended to stifle innovation. The most important reason is that people were still fairly spread out geographically. The 97 percent of people who lived outside cities, tended to live in isolated villages or hamlets. People living in such circumstances were rarely exposed to new ideas or technologies.

Diffusion of technologies, skills and social organizations within Agrarian societies was primarily vertical: from father to son and from mother to daughter. When children became adults, they supplemented this learning by copying other members of their local community. Because peasant villages were relatively low in population and each member was very similar to each other, there are relatively few other models to copy. There were simply too few people to come in contact with, so the diversity of options to copy was limited.

Members of Agrarian societies also had strong Survival-based values. By that, I mean that they were focused on group survival. Survival-based values tend to make members skeptical of new technologies, skills and social organizations. Because their lives hung on the edge of survival and they often face catastrophic disruptions such as plague, famine and wars, change was often regarded as bad and not worth the risk.

Members of Agrarian societies also had strong ethno-centrist beliefs. Because each ethnic group had a suite of technologies, skills and social organizations that changed relatively little for generations, they were invested with positive moral values and associated with their ethnic group.

The technologies, skills and social organizations of other ethnic groups were not just different, they were often regarded as strange and even immoral. Adopting objects or behavior that were associated with outsiders would make one stand out from the group and potentially be viewed as a traitor.

The political, economic, military and religious elites of Agrarian societies had a strong material incentive to reject innovations of technologies, skills and social organizations. Their standard of living and social status was based upon extracting wealth from the rest of society. New technologies, skills and social organizations potentially undermined their ability to do so. New ideas and behaviors might raise the expectations of the peasantry and create prosperous newcomers who might threaten elite domination.

In Agrarian societies, patronage (support from your superiors) was far more important than merit or achievement. The political, economic, military and religious elites usually lived in national or regional capitals. Those living in rural areas strove to find favor from those above them so they could move to regional capitals. Those living in regional capitals strove to find favor from those above them so they could move to the national capital. Those lucky enough to live in the national capital found that they still had to get by mainly upon political patronage from their superiors, rather than their skills and achievements.

Because wealth and income were so crucially tied to extraction by elites, merchants and artisans were forced to produce and market products that appealed to those elites. The only way to earn money was to produce and market luxury goods that provided amusement to elites and reinforced their social status. The peasantry simply did not have enough money to buy more food, tools and materials. Relatively poor artisans in villages and local market towns could purchase some goods and services, but there was rarely enough demand in these areas to create large markets.

Whereas today, we tend to look outward toward people who are similar to us, in Agrarian societies everyone looked upward: upward for approval, upward for markets, upward for patronage, upward for marriage. This meant that the tastes and preferences of the elite had a powerful effect on the rest of society.

While informal classes dominated Horticultural societies, elites within Agrarian societies formed impersonal institutions. Warriors were institutionalized into armies. Priests were institutionalized into churches. Royal followers were institutionalized into royal bureaucracy. Merchants were institutionalized into royal monopolies.

While today we live in a society with thousands of institutions — corporations, churches, charities, clubs, and relatively autonomous local governments — Agrarian institutions were far more limited in scope. In any one domain Agrarian elites believed that there must be only one institution, that institution must be highly centralized and it existed at the pleasure of the monarch or emperor.

Elites in Agrarian societies kept a very tight rein on the institutions that were allowed to exist. The king or monarch established royal monopolies in the political, economic, military and religious spheres. No new institutions were allowed to exist without royal permission. Inevitably, royal bureaucrats and their followers staffed these institutions and made sure that they followed the dictates of the sovereign.

New institutions with new ideas were expressly forbidden unless the monarch could be convinced it was for his benefit. Because new institutions play a key role in innovating new technologies and skills, these royal monopolies deadened innovation.

Institutions in Agrarian societies were overwhelmingly government-sponsored monopolies whose aim was to extract wealth to benefit of elites. Competition, diversity and decentralization represented potential threats to the established order, so they were not allowed.

These centralized, non-competitive monopolies tended to stifle innovation. Monopolies stamp out much of the variation that makes innovation possible. Except for the areas of military and elite consumption, Agrarian societies tended to experience diminishing rates of innovation over time.

The royal court was the center of every Agrarian society. The court consisted of the monarch, his advisors and an entire entourage of hangers-on. The court functioned not only as the political center; it also functioned as the economic, cultural, artistic and social center as well. This forced almost everyone who wanted to excel in a domain to focus

> **KEY INSIGHTS**
>
> All Agrarian societies relied predominantly on one staple food: The Middle East, Mediterranean and Europe relied mainly on wheat, while East, Southeast and South Asia relied mainly on rice.
>
> Because of the bounty from these simple grasses, Agrarian societies all had large, dense populations compared to other simpler societies.

a great deal of time on currying favor from the Royal Court. This meant that no one was ever truly independent. The opinion of the royal court always trumped competence.

While copying the successful is a highly effective strategy for diffusing innovations, in Agrarian society copying the successful often achieved the opposite. The successful in Agrarian societies won their success by birth or currying favor from elites. Because Agrarian elites were typically suspicious of innovation, copying the successful merely hardened the status quo. Other strategies for copying had the same results as there was simply too little diversity in society.

> **KEY INSIGHTS**
>
> Elites in Agrarian societies were obsessed with status. These elites created centralized institutions that acquired their wealth by extracting resources from peasants through taxation and land rents.

Agrarian transportation and communication technologies were not advanced enough to bridge significant geographic barriers. Therefore, face-to-face contact between people from different cultures was rare.

Most day-to-day transportation was by walking. Some regions with domesticated horses, oxen or camels adopted these animals for more rapid transportation, but land transport was still very slow. Carts and wagons were about as sophisticated a transportation option as the typical peasant could own.

Before the invention of moveable type and the printing press, there were also very limited options for distant peoples to communicate with each other. Books were rare and mainly religious in content.

Even after the invention of the modern printing press, the vast majority of people in most societies were illiterate. Most people had a difficult time even imagining the lifestyles of people in other societies, so there was no opportunity to copy them. All of these factors together seriously limited the rate of diffusion of technology, skill and social organization within Agrarian societies.

Zero-Sum Competition

A defining characteristic of Agrarian societies was that it was far easier to prosper by stealing other people's wealth than creating new wealth. Because the economy was overwhelmingly rural and additional income was often extracted by elites, economic growth was very slow, if it existed at all.

The opportunity for an individual to transform their wealth and social standing by work and innovation was very limited. For peasants in particular, life involved constant physical toil with little chance of improvement and a constant threat of immediate disaster from drought, war or disease.

KEY INSIGHTS

Agrarian societies used the extracted resources to build powerful militaries. These militaries conquered vast empires.

Agrarian elites distrusted innovation because it potentially threatened their power and status.

This depressing existence created a mentality among the inhabitants of Agrarian societies that was profoundly zero-sum: "I can only benefit at the expense of others; if someone else benefits, that hurts me." This zero-sum mentality undermined the voluntary cooperation that is essential for innovation and progress.

Wealth and income in Agrarian societies were based upon arable land. While the amount of arable land could be expanded at the expense of nature, it was usually necessary to take it forcefully from another person. Violence, war, land rents and taxes were the principal means by which to succeed. Hard work, skills and innovation were helpful, but either humans or nature could take the benefits gained away in a flash.

Agrarian societies had two different forms of competition. The first was the military competition between societies. Success in war was critical to all Agrarian societies. Victory meant increased wealth to extract. Failure in war meant diminished wealth and perhaps even extinction of the entire society. When they were not fighting a war, Agrarian societies were preparing to fight the next war.

The other form of competition in Agrarian societies was between individuals who sought to move up in the ranks of centralized extractive institutions. Because these institutions dominated society and they had huge resources in an otherwise poor society, moving up the ranks was the only realistic way to be successful (other than conquering new territories in war). So success in Agrarian societies was zero-sum in this way too.

> **KEY INSIGHTS**
>
> Agrarian societies had a zero-sum mentality: "I can only benefit at the expense of others; if someone else benefits, that hurts me."

The Legacy of Agrarian Societies

If we assume that a generation lasts 20 years and Agrarian societies emerged in 3000 BCE, by the year 1000 up to 200 generations had lived in an Agrarian order. Two hundred generations of people living where taking was more beneficial than creating had a profound effect on how people thought. People assumed that it was better to accept their fate and resist change, because change would likely be bad. Peaceful cooperation between strangers for mutual benefit did not play a role in how these people saw the world.

While the legacy of previously discussed Society Types towards the promotion of progress was quite negative, the legacy of Agrarian societies is more mixed. It is striking that none of the early leaders in innovation and progress (Northern Italy, Flanders, the Netherlands, Southeast England and the United States) were Agrarian in the year 1500. Agrarian societies, while they inherited a large base of technology and large cities, tended to stifle innovation.

On the other hand, it is also interesting how many of the nations that have been able to industrialize within the last 150 years were previously Agrarian societies. These include Germany, France, Russia, Japan, South Korea, China, India as well as large swathes of the rest of Europe and Asia.

While no society or minority that is descendant from Hunter-Gatherer, Fishing, Herding or Horticultural societies in the year 1500 have been able to achieve rapid growth in wealth without the aid of extracting fossil fuels, dozens of Agrarian societies have been able to do so. And immigrants from societies that were Agrarian in 1500 have been able to prosper throughout the globe (Chinese, Indians, Jews, Lebanese, Germans and Japanese).

Herding Societies

In some natural environments not conducive to large-scale plow-based agriculture, a new society type emerged around 1000 BCE: Herding societies. Whereas Agrarian and Horticultural societies subsisted largely on growing domesticated plants, Herding societies subsisted on herding domesticated mammals on the open range (camels in the Desert; Reindeer and Yaks in the Tundra and Boreal forests; Horses in Temperate Grasslands; Cows in the Savanna; Sheep and Goats in the Temperate Highlands).

Key technologies enabled Herding societies to do the following:

- Domesticate wild herd animals (particularly the horse and camel)
- Ride horses and camels (using the saddle, stirrup, halter, bit and bridle)
- Transport goods rapidly across land (using horse-drawn carts and wagons)
- Wage war against other societies (using chariots, horse riding and firing composite bows)

As they lived in hostile environments without defensible frontiers, the members of Herding societies often evolved into fearsome warriors. The skills necessary for herding and slaughtering large animals translated well into the skills needed for ambushing and killing humans. In particular, the ability to use the composite bow-and-arrow from a galloping horse gave Herding societies a critical military edge.

Important Herding societies in history have included the Mongols, Iranians, Arabs and Turkic peoples. Together they inhabited a broad swath of Eurasia stretching from Mongolia, through Central Asia, Iran, the Middle East and North Africa. Because these regions are contiguous with most of the regions that supported Agrarian societies, these Herding peoples have played an outsized role in human history.

While Herding societies were poorer and much smaller in population size, they were man-for-man far superior to Agrarian societies militarily. In their home region, they could easily avoid combat when outnumbered and could rapidly concentrate where their opponents were weakest.

And where they had horses and camels, Herding societies could conquer vast Agrarian empires. While they could not absorb the far larger populations of Agrarian societies, Herding societies established small warrior elites that ruled the cities and farmers through fear. Because Herding societies were fundamentally unsuited to farming, however, these empires were usually ephemeral.

By far the most historically significant Herding societies lived in the Central Asian steppe. In the period between 400 and 1500, these peoples formed expansive empires that dominated the Agrarian societies of China, Korea, India, Persia, Russia and the Middle East. While these empires were often short-lived and led to few long-term changes, they did have a devastating impact on the lands that they conquered.

Despite their ability to dominate Agrarian societies, it is doubtful whether Herding societies were able to deliver progress to their people in the long term. Undoubtedly, the conquest of richer Agrarian societies enabled them to extract greater resources from farmers, but most of these empires were fleeting in duration.

More importantly, these Herding empires were parasitic in nature. Rather than creating wealth, they extracted wealth from Agrarian societies through military conquest. Once Herding societies lost their military dominance because of the invention of firearms, their people returned to their poorer Herding lifestyle.

The Legacy of Herding Societies

The legacy of Herding societies is still with us today. The ancestors of Herding societies in the year 1500 — Central Asia, the deserts of the

Middle East and North Africa, the Savanna of Africa and the Tundra of the far North — have had trouble copying the success of modern Industrial societies. To the best of my knowledge, no society or sub-national minority whose ancestors descended from Herding Societies in the year 1500 is above average in social-economic success.

While some were able to achieve wealth through the extraction of fossil fuels and exporting them to wealthier Industrial societies, the ancestors of Herding societies have had difficulty achieving success without this somewhat artificial resource. And the enormous wealth derived from fossil fuels has enriched only a small elite within those countries.

Free Peasant Societies

There is one more type of society that should be briefly discussed: the Free Peasant society. Switzerland, Denmark, Norway and Sweden were the most important Free Peasant societies in history. Early Flanders (modern-day Belgium) and the Netherlands were also Free Peasant societies in their early development.

Free Peasant societies were extremely rare and had a limited impact on world history. Free Peasant societies shared the subsistence patterns of Agrarian societies and their lack of large trade-based cities. But Free Peasant societies lacked the centralized extractive institutions of Agrarian societies. Free Peasant societies were an ocean of peasants who typically owned their own land or rented it consensually from landlords.

While Agrarian societies glorified status and hierarchy, Free Peasant societies were fiercely egalitarian. They lacked titled nobility, or where they existed, titled nobilities owned a very small percentage of the land.

KEY INSIGHTS

Herding societies survive by herding domesticated animals on the open range. Their key technologies are the saddle, carts, wagons, chariots, and composite bow.

Herding societies often evolved into fearsome warriors, capable of conquering vast empires of wealthier people. Rather than creating wealth, they extracted wealth from agricultural societies through military conquest.

Those titled nobles were also not highly influential over political affairs.

As we will see later, the lack of centralized extractive institutions made Free Peasant societies much more willing to adopt technologies, skills and social organizations from more economically successful societies. The lack of centralized extractive institutions enabled Free Peasant societies in Europe to industrialize much earlier than Agrarian societies.

It is not an accident that these societies are still some of the most egalitarian societies on the planet. Nor is it an accident that these societies have far higher levels of trust than most other peoples. Their cultures evolved without the domination of extracting elites, so peasants learned to trust each other and work together to solve common problems in highly efficient ways. Because of all of this, societies that were once poor Free Peasant societies are now some of the wealthiest and most prosperous societies on Earth.

Conclusion

The history of humanity up until quite recently is rather depressing. Though there was an enormous amount of innovation in technologies, skills and social organizations, very little of the derived benefits went to the masses. The vast majority of humanity engaged in an endless physical struggle to produce enough food to eat. Virtually all the benefits of that struggle went towards growing a larger population and enriching extractive elites.

The main sources of short-term change were negative: war, famines, drought and epidemics. Unless someone was willing and able to take from others, there was little chance of improving oneself within their lifespan. If someone was lucky enough to avoid wars, famines, droughts and epidemics, they might be able to live a decent life, but it would be nothing like the kind of progress that we take for granted today.

Such has been the bulk of the human experience.

GEOGRAPHICAL CONSTRAINTS TO PROGRESS

So far we have seen that throughout history societies experienced very little progress. We have also seen that radically different types of societies evolved. Each new society type gradually increased the rate of innovation of technologies, skills and social organizations. But many regions were entirely left out of this change, persisting in the same society type for millennia.

Despite the progress of recent centuries, there are still enormous inequalities in how people live their lives. Today some regions have industries churning out complex digital and industrial technologies, while in other remote areas humans still practice hunting and gathering techniques for acquiring their food. The wealthiest societies are tightly clustered into a small portion of planet Earth.

Why are there such contrasts between nations? If progress has been so widespread, why don't all nations have the same standard of living as the United States and Denmark? What accounts for these radical differences?

While many theorists try to account for these differences with contemporary causes — institutions, political leaders, government policies, trade, colonialism, exploitation, and racism — these differences go far back into history.

For example, if one compares a rank-order list of how prosperous a society is today with the same list for the year 1000, one can see a great deal of commonality. Societies in Europe and those that had recently been settled by Europeans, the Middle East, South Asia and East Asia rank relatively high, while societies in South America, Sub-Saharan Africa, Siberia, New Guinea, Australia and the Pacific Islands rank relatively low.

The differences go back even further than the year 1000. In a fascinating article entitled "Was the Wealth of Nations Determined in 1000 BC?", William Easterly and other researchers find a strong correlation between levels of technology and per capita income today with levels in the years 1500, 1000, 0 and even 1000 BCE!

Another study by Louis Putterman and David Weil found that 46 percent of the current variation in per capita income is determined by whether a nation's people are genetically descended from Europe (even after accounting for other factors). Other researchers have claimed that the time since the society adopted agriculture is a dominant factor in explaining a society's political development.

Many other studies have found similar results: current differences in wealth and levels of technology are strongly associated with factors in the same society thousands of years ago. This burgeoning literature clearly shows that we cannot account for differences in the wealth of societies with current factors or even factors that are decades or centuries old. We must look back millennia.

One of the problems with determining ultimate causes is that it is always possible to go further back. If one determines, for example, that

> **KEY INSIGHTS**
>
> Inequalities between societies are not new. They go back as far as 3000 years ago! So current inequalities are not caused by current factors, such as institutions, political leaders, government policies, trade, colonialism, exploitation, or racism.
>
> Inequalities between societies today are caused by geography.

sailing ships caused important differences in economic development, one can then ask "but what caused sailing ships to be invented in the first place?" If we take this line of questioning to the extreme, then everything is ultimately caused by the Big Bang!

Obviously, using the Big Bang as the ultimate cause that explains human history is going too far, so we need a way to push back the explanation far enough to avoid a long chain of causality, while still not constraining one's explanatory variables to the present day.

For this reason, I believe that it is necessary to start with causal factors that were in place before modern humans evolved. If such a factor can be used to explain a great deal in the variation of wealth in human societies, then we have gone far enough back that we can be confident that we have found an ultimate cause.

In previous chapters, I explained the role of biology, energy and evolution. All three of these factors preceded the evolution of modern humans. But these are all factors that human societies had in common, so we need another factor to explain differences between societies.

While the answer to why some societies are much wealthier than others is complex and I wish to fully answer the question in later books, I believe that geography is the dominant cause. But geography doesn't directly place constraints on the development of today's society. The impact of geography is more indirect than that.

Geography placed hard constraints on the ability of humans to acquire food. The type and amount of food that could be produced either constrained or promoted different society types (as discussed in the previous chapter). Those society types, in turn, either promoted or

---

KEY INSIGHTS

Geography placed hard constraints on the ability of humans to acquire food. The type and amount of food that could be produced either constrained or promoted the type of society that could evolve. Those society types, in turn, either promoted or constrained the innovation of certain technologies, skills, social organizations and values.

Geographical constraints created a poverty trap that persisted for millennia.

constrained the innovation of certain technologies, skills, social organizations and values. Without key technologies, skills, social organizations and values, people living in certain societies were not able to innovate and copy the innovations of more complex societies.

Geographical constraints created a trap that persisted for millennia. This created a huge gap in the standard of living between societies and meant that some societies enjoyed progress for centuries while others stagnated until relatively recently.

This chapter will go into more detail as to which forms of geography promoted and hindered the development of complex societies. Only in a small sub-set of those relatively complex societies was progress possible.

Biomes

Our quest to find geographical factors that promote and hinder progress must start with biomes. Susan Woodward's eight-book series *Greenwood Guides to Biomes of the World* has done much to extend our knowledge on this subject.

A biome is a category for a geographical area based upon its dominant vegetation. The dominant vegetation in any area sets very broad constraints on what type of plants and animals can evolve there. The dominant vegetation sustains the local herbivores, which then sustain the local carnivores. Since humans use a portion of these plants and animals as base materials for food production, biomes also play a critical role in constraining the types of human societies that can develop within them.

The concept of biome has recently been systematized by the World Wildlife Fund (WWF), an organization dedicated to understanding the diversity of life on planet Earth. The WWF's system divides the earth into 14 different biomes containing 867 smaller ecoregions. Biomes are determined by many geophysical and meteorological factors, including latitude, elevation, ocean currents, prevailing wind, rainfall and seasonality.

A full-color version of this map can be found at my website. See the front matter of this book for a URL and a QR code to the website.

One example of a biome is the Temperate Grassland biome. Not surprisingly, grasses dominate this biome, and the biome is located within the temperate latitudes. The California Central Valley and Mongolian-Manchurian grassland are ecoregions. Each of these ecoregions is an instance of the Temperate Grasslands biome. While these two ecoregions differ from each other, they share far more in common.

The WWF biome categorization is an internationally recognized system for understanding wild plants, animals and their ecosystems. Unfortunately, those who study human societies have not used systematic analysis of biomes.

I believe that this is a big oversight by social scientists and historians. The fact that the WWF system was developed without reference to human societies means that it gives us a relatively objective view of geography; one that can be used to look for patterns in human development.

When one compares the WWF map to human population density maps for any given point in history, one can discern clear patterns. Very few of the biomes have been able to support large population densities until recently, and even fewer have been able to support the complex societies that drive innovation and progress.

Biomes That Made Progress Impossible In The Past

First of all, no large permanent human settlements have ever existed in any of the **Ocean** biomes. Since oceans cover about 70 percent of the planet, this eliminates a huge portion of planet Earth from human habitation.

Even with 21st Century technologies, our interactions with the ocean consist entirely of short-term voyages for transport and fishing. While the ocean provides resources for many land-dwellers who are brave enough to voyage upon it, it is more hostile to human habitation than even the driest deserts.

The ice-covered **Polar regions** and the **Highlands regions** over 12,000 feet in altitude are also almost devoid of human habitation. Eskimos and Inuits have learned how to live in the polar biome, but their population sizes are very small. Both cultures have developed Fishing societies that subsist primarily by hunting marine mammals, such as seals and whales. Even these highly specialized cultures have only been able to survive on the edge of Polar regions.

While modern scientists, explorers and adventure enthusiasts sporadically visit them, these regions are not much more hospitable than the ocean. It is clear the Polar regions and mountains over 12,000 feet in altitude cannot create a food surplus to support high population densities, cities and rapid technological innovation. Even the simplest forms of agriculture are impossible.

Somewhat more hospitable are **Tundra** biomes. Tundra biomes are largely confined to the very northern edge of Asia and North America. The dominant vegetation in the Tundra biome is ground-hugging vascular plants, mosses and lichen. The only large grazing animals are caribou/reindeer and musk ox.

Temperatures on the Tundra are above freezing for only 6-10 weeks per year, causing a very short growing season. Even worse, just a few inches below the surface is a thick layer of permafrost, which is essentially frozen soil. This permafrost layer makes it impossible for tree roots to anchor in the soil.

The result is a biome that cannot support trees. This creates a major constraint on human habitation. Without trees, wood cannot be used for the construction material, tools, heating of homes or smelting metals. And without trees, there is no natural shelter from the dry arctic winds, so even bushes cannot survive.

Because of the very limited plant growth in the Tundra biome, only two types of human societies have been able to survive there: Fishing societies that subsist on marine mammals (similar to Polar regions) and Herding societies that subsist by herding reindeer. Not surprisingly, these types of societies cannot generate food surpluses that support the high human population densities needed for innovation and progress.

> **KEY INSIGHTS**
>
> Biomes that made progress impossible are the Polar, Highlands, Tundra, Boreal Forest, Temperate Grasslands, Desert, Savanna and Tropical Forest biomes. Each made it impossible for humans to achieve the first key to progress: productive agriculture.

Progressing southward through the Northern Hemisphere, we come to the **Boreal Forest** biome of Siberia, Canada and Northern Scandinavia. While most Boreal ecoregions are in the far north, some extend southward along mountain ranges.

Regardless of location, the winter temperatures in Boreal Forests are very cold, and the summer growing season is very short. Up to six months per year see below-freezing temperatures. While the biome experiences 15-20 inches of precipitation per year, most of that precipitation falls in the summer as rain.

Conifers dominate the Boreal Forest biome. Conifers are trees covered by needles, rather than leaves. Whereas tropical biomes have a huge variety of plants, enormous tracts of Boreal Forest are typically covered by only two or three species (usually a variety of pine, spruce, fir or larch).

In very cold climates with short growing seasons, needles on conifer trees offer many advantages over leaves. The dark green color of most needles absorbs a wider spectrum of wavelengths of sunlight, while their waxy coating traps scarce water inside the plant. Conifers can also start producing sugar via photosynthesis as soon as the late northern spring starts without having to devote energy towards regrowing leaves. Finally, their conical shape helps to shed the deep winter snow.

The biggest disadvantage of conifer trees for other species, including humans, is that needles decay very slowly. The combination of thick waxy needles and a very limited amount of decomposing bacteria

Biomes

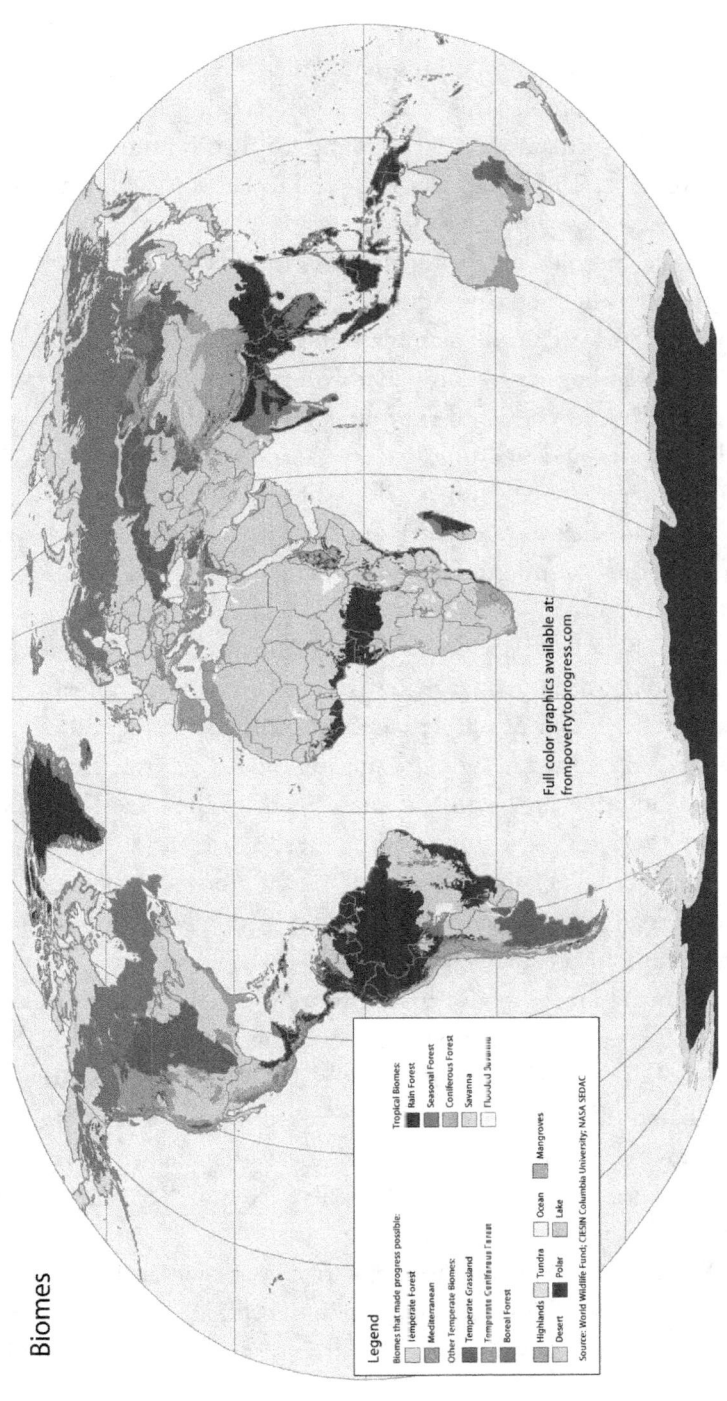

Legend

Biomes that made progress possible:
Temperate Forest
Mediterranean

Other Temperate Biomes:
Temperate Grassland
Temperate Coniferous Forest
Boreal Forest

Highlands   Tundra
Desert   Polar

Ocean   Mangroves
Lake

Tropical Biomes:
Rain Forest
Seasonal Forest
Coniferous Forest
Savanna
Flooded Savanna

Source: World Wildlife Fund; CIESIN Columbia University; NASA SEDAC

Full color graphics available at:
frompovertytoprogress.com

and animals in the region means that a thick layer of dried needles usually covers the topsoil. This interferes with seedlings' growth, creating "dead zones" with at most a few saplings or shrubs immediately below each large tree. These mats cause the soil to become highly acidic and unproductive for plants or agriculture.

Conifer needles, except for the buds that bloom in the spring, are also not very nutritious for animals. Therefore, very few boreal animals eat needles. Most Boreal mammals and birds subsist on seeds, nuts, leafy plants in bogs or by eating herbivores.

Boreal forests consist of enormous stands of almost identical trees. Pools of water and lightning strikes cause the only real variation in Boreal regions: the landscape is dotted by a myriad of burn sites, rivers, bogs and lakes in the spring and summer. These open, wet regions create Tundra-like micro-regions that support leafy plants during the spring and summer.

Hunter-Gatherer societies and a few isolated Fishing societies are the only types of human societies that have inhabited Boreal regions. Because of the lack of edible plants and unproductive soil, societies in this biome subsist heavily on hunting moose and trapping weasels, beavers and other fur-bearing animals. None of these food sources can support the large dense populations needed to promote innovation and progress. Just like the Polar and Tundra biomes, the Boreal Forest biome sets a hard ceiling on the level of development that humans can achieve without outside help.

The next biome on our trip southward is the **Temperate Grassland** biome. The bulk of this biome is located in Central Asia and North America, but smaller Temperate Grassland ecoregions exist in Argentina, Australia and the Middle East. As its name implies, grasses and forbs dominate the Temperate Grassland biome. Other characteristics of this biome include a semi-arid climate with 10-20 inches of rain per year, a cold winter and a hot summer.

While trees find this combination of climatic factors hostile, grasses thrive in it. Ten million years ago, when the climate was much warmer, grasslands did not exist in large enough concentrations to be considered a biome. But as the climate gradually cooled and dried, grasses began to establish local footholds. About 4-6 million years ago, conditions were sufficiently cool for grass biomes to evolve. The tough, hardy grass grew rapidly at the expense of tropical forests.

Not by coincidence, this is exactly the period when our primate ancestors first ventured out of the trees and moved into the grasslands. Without the evolution of Grassland biomes, it seems very unlikely that humans would have existed at all.

The key characteristic that distinguishes grasses from trees and bushes is that the bulk of their organic content is in their roots below the soil. While trees and most other large plants grow upward to access as much of the sun as possible, grasses grow downward in their roots. This makes grass extremely resistant to cold, fire and grazing. While parts of grass may be destroyed above the surface, their deep root system traps moisture and uses that moisture for regrowth.

It seems likely that fire has been the most potent ally of grass in its endless battle with trees. Random lightning strikes or even fires deliberately created by humans would have had a devastating effect on trees, but little effect on the largely underground grass.

Each tree that dies would be quickly swarmed by new grass shoots, which would grow a dense network of roots so fast as to starve any tree saplings that might attempt to grow back. In this way, grasses living in dry climates have gradually advanced at the expense of trees.

Temperate Grasslands have low levels of plant diversity; they are often covered by hundreds of miles of only a few species of grass. Whichever types of grass are best adapted for a locale usually grow so thick that they choke out most other plants. This in turn forces all local herbivores to evolve to digest those species efficiently.

Temperate Grassland animals consist of grazing herbivores and the carnivores that hunt them. While this leads to low levels of animal diversity, the biome supported the native wild ancestors of the domesticated animals necessary to support agricultural societies.

Just as important, the grasses themselves evolved into the ancestors of staple crops that form the bedrock of agricultural societies: rice, wheat, oats, rye, millet and others. Grasses have been the biological foundations upon which complex societies have been built.

With all of these biological advantages, one might think that Temperate Grasslands could support advanced Agrarian societies. Unfortunately, grass has one huge disadvantage for humans: their dense mat of roots is so thick and deep as to be almost impossible to plow.

While the soil is usually extremely productive, it is so hard to work that most farmers give up and move elsewhere. Until the invention of the steel plow in the 1830s, most agriculture in the Temperate Grassland biome was restricted to isolated regions where the sod was naturally broken by riverbeds.

The dominant type of human society in the Temperate Grasslands biome has been the Herding society. Humans in this society type subsist by herding large herds of herbivores: horses, cattle, sheep and goats. Those animals enabled humans to exploit the largest energy source of the grassland: cellulose. By eating animals that could digest cellulose, the bulk of grass material, humans were able to live in an otherwise hostile environment.

The horse has been the most important of all herd animals to human development. Horses have provided three critical resources for human societies. At first, horses functioned mainly as a food source, not that different from cattle. Early Herding societies probably ate horsemeat. More importantly, horses provided the most rapid form of transportation in the pre-Industrial era. Finally, Agrarian societies learned how to harness horsepower to pull plows, carts and wagons.

But ultimately, the Temperate Grasslands biome placed a hard ceiling on the type of human societies that could develop. Where large domesticable animals existed, as in Asia, Herding societies could evolve. Where large domesticable animals did not exist, as in North America, South America, and Australia, humans could not evolve beyond Hunter-Gatherers. Regardless, productive agriculture capable of supporting densely populated cities was not possible until after 1830. This ultimately undermined the amount of human progress that could take place in these regions until that date.

Another biome that was historically dominated by Herding societies is the **Desert** biome. Deserts are defined by annual precipitation of fewer than 10 inches per year. With so little moisture, trees have a hard time living, so smaller shrubs dominate. Desert shrubs usually have very small leaves to preserve moisture, and many are covered in spines or thorns. These shrubs have shallow but wide root systems to capture infrequent rains and long, deep taproots to gather groundwater deep under the surface. Many other desert flowers lie in seed until rainfall is high, causing a spectacular desert spring.

The only Desert plants that are useful to humans are perennial forbs that store their moisture and nutrients in bulbs and rhizomes. These forbs are much sought after by Hunter-Gatherer peoples in the Desert biome.

The lack of water and hot temperatures make it extremely hard for mammals to live in the desert. The desert is reptile country, as these animals can use the sun to warm their bodies and their slow metabolism means that they can survive without eating for long periods.

One species of mammal did find a way to survive and prosper in this hostile environment and became a cornerstone for Desert Herding societies: the camel. With their ability to pack heavy loads across long distances without regularly drinking water, camels have been indispensable for long-term survival in Desert ecoregions.

The only Agrarian civilizations in Desert biomes were those that evolved on large rivers in Southwest and Central Asia. Those societies exploited this unusual niche to grow vast amounts of wheat, rye, millet and other grains. Those grains enabled some of the earliest Agrarian societies to evolve: Egypt, Mesopotamia, the Indus river valley and scattered cities throughout Central Asia. But the riverine nature of their societies placed hard geographical constraints that meant they were weaker than Agrarian societies in other biomes.

A third biome that has been dominated by Herding societies for centuries is the **Savanna** biome (or Tropical Grasslands). Africa has the most famous Savanna ecosystems, but South America and Australia also have substantial ecoregions of that type. These regions are in tropical latitudes that experience less than five inches of rain for at least five months per year. Just as in the Temperate Grasslands, this level of rainfall creates a perfect habitat for grass.

Unlike the Temperate Grasslands biome, the Savanna supports a large variety of herbivores. Each species has evolved its own specialized means of consuming grass: some prefer different heights or species of grass; some prefer grazing at a specific time of day or time of year. And the large predators of the Savanna have also evolved specialized means for hunting these grazers.

With all of these large herbivores, one might think that there were plenty of options for domestication. Unfortunately, Savanna herbivores are often aggressive and ill-tempered, making domestication impossible. Horses, cows, pigs, sheep, goats and other domesticated mammals all originated in Temperate latitudes.

This created a major barrier to the evolution of both Herding and Agricultural societies on the Savanna. Not until cattle diffused from the Nile River valley were Herding societies able to evolve on the African savanna. Because they lacked domesticable animals, the Savanna ecosystems of South America and Australia did not evolve beyond Hunter-Gatherer societies until recently.

About 50 million years ago, when the planet was significantly warmer, the **Tropical Rain Forest** biome dominated planet Earth. Flowering plants, the dominant vegetation in Tropical Forests first evolved during the Cretaceous, the last period inhabited by dinosaurs. Some scientists believe that it is was the emergence of flowering plants that led to the tremendous variety of dinosaur species during that period. The most famous were Triceratops and Tyrannosaurus Rex.

The Chicxulub meteor impact, which drove the dinosaurs to extinction and enabled mammals to become the dominant animal form, also enabled flowering plants to largely replace the conifers that had dominated during the dinosaur era. The gradual cooling and drying of the climate over the last 50 million years caused Tropical Forests to contract numerous times, including during modern times.

Today Tropical Rain Forests cluster very close to the equator at elevations below 1000 meters. The bulk of the ecoregions of this type is in Southeast Asia, Amazonia and the Congo, although smaller ecoregions exist in West Africa, India, Central America and Southern Brazil.

Huge broadleaf evergreen trees that can grow to a height of 100 feet dominate the Tropical Rain Forest biome. Below them exist at least four more layers of canopy. The floor of these forests is usually open ground, due to the lack of sunshine on it. Without this sunshine, few plants can grow. This means most food sources are high up in the trees — most inconvenient for ground-dwelling humans.

Unlike all other biomes, there is very little variation in heat, moisture, sun angle or day length from day-to-day or even season-to-season. This means that each plant species was free to evolve a flowering and fruiting season at widely differing times of the year.

Strangely, the key constraining resource in the Tropical Rain Forest is sunlight, because the tallest trees block the sun from hitting other shorter plants. This causes plants to compete against each other for sunlight by growing upward.

Tropical Rain Forests are a biome with a staggering variety of plant and animal species. While Temperate biomes tend to have a small number of species over a large land area, one can find hundreds of plant and animal species within a single square mile in Tropical Rain Forests.

One might think that, given the number of wild plant species, humans would have a cornucopia of domesticated plants. While there are many native wild ancestors of domesticable plants in the Tropical Rain Forest, staple crops, which tend to be grasses, are notably lacking. Humans in the Tropical Rain Forest mainly subsist on root vegetables: cassava, aroids, sweet potatoes and yams. These crops are far more difficult to preserve and transport than the grains grown in temperate latitudes.

Even worse, Tropical Rain Forests tend to have unproductive soils. The nutrients are trapped in leaves that never fall to the ground. The high heat and moisture have leached most of the nutrients from the soil. The high number of decomposing species in the soil means that the nutrients from fallen trees and shrubs do not penetrate into the soil.

Unfortunately for humans, most of the animal species in Tropical Rain Forests spend all or almost all of their time high up in the trees to gather wild fruits and avoid predators on the ground. Until the invention of the bow-and-arrow, hunting in the jungle was a poor method of supporting one's family. For this reason, human societies in the Tropical Rain Forest generally subsist on root vegetables and fish with only small amounts of calories coming from hunting.

While the Tropical Rain Forest is a rich haven of biological diversity, it places serious constraints on the types of human societies that can evolve. Only Hunter-Gatherer, Fishing and Horticultural societies have been able to evolve in this biome. None of them are capable of supporting the highly productive agriculture needed to create densely populated cities.

Closely related is the **Tropical Seasonal Forest** biome. This biome shares much in common with the better-known Tropical Rain Forests, with but one very important difference. While Rain Forests experience high rainfall year-round, Seasonal Forests have a distinct Dry season, usually for four to seven months.

While Rain Forests hug the equator, Seasonal Forests generally live between latitudes 10 and 25 (i.e. just north and south of the Rain Forests). The largest ecoregions of this type are located in Indochina, India, Central America and scattered areas in South America.

While Rain Forests typically have over four canopy layers, Seasonal Forests typically have only one or two layers. Trees in Seasonal Forests tend to be shorter with smaller leaves and many are covered in thorns. Many species are deciduous, so their leaves nourish the soil.

In terms of their impact on the evolution of human societies, Seasonal Forests are quite similar to Rain Forests. Except where there are fertile river deltas, Agrarian societies cannot evolve in this biome. Elsewhere, human development is restricted to Hunter-Gatherer, Fishing and Horticultural societies.

Biomes That Made Progress Possible

By far the most conducive biome for the evolution of complex human societies is the **Temperate Forest** biome. The Temperate Forest biome evolved on temperate latitudes near oceanic currents that carry warm water from tropical latitudes. These warm water currents create the conditions for annual rainfall of over 20 inches per year. The result is a relatively warm, wet climate that is perfect for deciduous trees.

Northern Europe has the single largest Temperate Forest ecoregion, although East Asia and eastern North America also have very large ecoregions of this type. Chile, Australia and New Zealand each have small Temperate Forest ecoregions. Temperate Forest biomes typically have growing seasons of around six months, with even longer growing seasons in the south.

The powerhouses of the Temperate Forest biome are the leaves on deciduous trees, most of which are species of beech, walnut, ash, maple or birch. While needles on conifers and leaves in tropical latitudes stay on the tree year-round, most deciduous trees in temperate latitudes go through a seasonal cycle of losing their leaves in the autumn and then regrowing them in the spring.

Losing leaves in the autumn protect the branches from cold, dry winters; growing them back in the spring enables trees to deploy what are effectively organic solar panels. Leaves are typically very thin and wide, giving the tree a huge amount of surface area for catching sunshine for photosynthesis. While the trees have evolved this behavior for their survival, the benefits to the rest of the ecosystem are impressive.

Leaves are packed full of sugars and other nutrients, perfect for consumption by insects and browsing mammals. When nitrogen-rich leaves fall to the ground in the autumn and then unfreeze in late winter and early spring, they create a thick layer of hummus on top of the soil. Spring rains wash the nutrients deeper into the soil, where huge numbers of decomposing bacteria and animals ensure these nutrients nourish the deepest layers of soil. The thick hummus also provides a perfect habitat for earthworms and burrowing mammals, which aerate the soil.

During the winter, the trees and many animals in Temperate Forests go into dormancy. Birds usually migrate to warmer climates. The next spring when the snow and ice melt, those nutrients from the leaves are then gathered by plant roots from the soil and redirected into growing new leaves.

Because fresh young leaves contain the most nutrients, spring in Temperate Forest biomes is an explosion of life. Insect larvae hatch to take advantage of this bounty, luring in migratory birds for the feast. Browsing mammals give birth to babies for the same reason. This constant seasonal cycle ensures that Temperate Forests have some of the most productive soils in the world.

KEY INSIGHTS

Progress was possible, but not inevitable, in the following biomes: Temperate Forest and Mediterranean biomes. Both made it possible for humans to achieve the first key to progress: productive agriculture.

In addition, trees themselves give humans one of the most useful materials in the world: wood. Wood can be used for heating during the night and winter, a necessity in colder temperatures. Wood is also one of the most useful materials for the construction of tools, dwellings and transportation devices. Wood can also be used to make charcoal, a crucial first step in metalworking technology.

Humans gradually learned how to use wood, rain and fertile soil to create productive agricultural systems capable of generating a food surplus. This food surplus enabled dense human populations that promote innovation and progress. Countries situated in the Temperate Forest biome include China, Japan, Korea, USA, UK, Netherlands, France, Germany, Russia and Northern Italy. Citizens from societies inhabiting the Temperate Forest biome have recorded the bulk of written human history.

The one key disadvantage of the Temperate Forest biome is the relative lack of native wild ancestors of domesticable plants and animals. Virtually all such species in use today arrived via human migration from other biomes.

Horses, cattle, sheep, goats and pigs all originated in Temperate Grassland biomes in Asia. For this reason, to have a big impact on human history, the first successful societies located in Temperate Forest biomes were in Eurasia.

Geographical isolation from Eurasia is the key reason why it took so long for complex societies to evolve in Temperate Forest biomes of North America, Chile, Australia and New Zealand. These regions lacked domesticable animals, and they were not near regions that possessed them. The arrival of immigrants from complex societies in Europe triggered some of the fastest societal changes in human history because this missing ingredient suddenly arrived.

The other biome that is capable of supporting productive agricultural systems is the **Mediterranean** biome. The Mediterranean biome evolves on the western side of continents between the latitudes of 30 and 40. Its key distinguishing characteristics are rainy winters and a long, hot and dry summer. This rare combination creates a short growing season when temperatures are mild and the soil is moist.

Not surprisingly, given its name, the bulk of the Mediterranean biome is located along the Mediterranean Sea in Europe, North Africa and the Levant. Smaller and more isolated Mediterranean ecoregions can be found in Chile, California, South Africa and Australia.

The Mediterranean biome is for all practical purposes a desert for nine months out of the year. For this reason, most of the plants have evolved to conserve water through the long dry season. Most plants are evergreens with thick leathery leaves, and many of them protect themselves from herbivores with aromatic oils that humans use in cooking Mediterranean cuisine.

While not as conducive to the evolution of complex societies as is the Temperate Forest biome, the Mediterranean biome has supported many civilizations, particularly Ancient Greece, Carthage and Rome. Its primary disadvantage is the lack of an annual leaf fall, so the soil tends to be much less productive. And fewer trees mean that wood is not as plentiful.

So we can see that biomes placed enormous restrictions on the types of human societies that could evolve. Most biomes are simply incapable of supporting productive agriculture, the first Key to Progress. Some biomes are only capable of doing so where there are rivers, a very small portion of their landmass.

Only the Temperate Forest and Mediterranean biomes are capable of supporting vast expanses of productive agriculture using pre-Industrial technology. And many regions with a Temperate Forest or Mediterranean biome lack domesticable animals that make plow-based agriculture possible.

Rivers

Many of history's most important cities have been located on the banks of great rivers. While rivers are not essential for the development of complex Agrarian cities, they do make them far more likely. If a region has other geographical factors that promote the development of complex Agrarian societies and there is a river within that area, a large city will likely evolve on its banks.

Rivers offer crucial advantages to human development; they:

- Offer sources of (hopefully) clean drinking water.
- Make it easier to remove human waste.
- Offer sources of irrigation for crops.
- Deposit additional nutrients in the soil for growing crops.
- Enable cost-effective transportation of people and freight.
- Offer defensible lines to block the approach of enemy armies.

While rivers are common in many regions, not all rivers are useful for Agrarian societies. In most cases, the river must be located in a biome that enables plow-based agriculture (as mentioned above). In some cases, as in the Desert and Seasonal Tropical Forest biomes, rivers can effectively create little islands of agriculture within an otherwise hostile biome.

In addition to being located in a hospitable biome, to support complex societies, rivers must have all or almost all of the following characteristics; they must be:

- Substantial in width, so that enough water is carried.
- Navigable by large boats (to facilitate trade)
- In flow all year-round, particularly near the mouth
- Emptying into an ice-free ocean year-round

- Free from large cataracts or waterfalls (except near their upstream source)
- Free from large marshes near their mouth.

To fulfill all of these qualifications, a long section of river must be located on gently sloping terrain, but not so gently sloping as to allow the formation of marshes. Many rivers do not have long stretches such as this, a key barrier to the development of complex Agrarian and Commercial societies.

Altitude

While biomes and rivers are the most important geographical constraints on the development of human societies, there are others. As Joel Cohen and Christopher Small have shown, altitude is also an important constraining factor. While landmass on Earth varies in altitude from just below sea level to 10,000 meters above, about 73.7 percent of humanity inhabits altitudes of less than 500 meters above sea level.

Plains offer many advantages over highlands, particularly in Temperate latitudes. Soil tends to concentrate on coastal plains after having been carried down from the mountains by runoff from rain and snowmelt. The steeper the grade on the mountainside, the more denuded the soil on the slopes. Some of that soil is deposited in mountain valleys, but given that most valleys are relatively small in size, flat plains below mountains tend to have far deeper topsoil.

Mountains also create substantial barriers to transportation. While horses, carts and wagons can range widely over the plains, they have an extremely difficult time on steep mountains. For this reason, transportation in mountainous regions is often constrained to valleys and passes.

In tropical latitudes, altitude has some counter-veiling advantages. Plains in Tropical Forest regions are often death traps to those without biological immunities, with endemic malaria, dengue, sleeping sickness, tsetse flies and other debilitating diseases.

For this reason, tropical populations in forested regions tend to be concentrated in altitudes between 1500 to 3500 meters. Central Mexico, the Andes and South-Central Asia are the most important examples of relatively dense human populations at high altitudes.

Soil

Because agriculture is key to feeding dense human populations, the type of soil in a region is critical to its ability to support complex societies. Soil is the combination of weathered rock, air, water and organic matter from decomposing plants and animals.

KEY INSIGHTS

The vast majority of humanity inhabits altitudes of less than 500 meters above sea level. These people are also richer than those living at higher altitudes.

Because the dominant vegetation makes up the bulk of the organic matter, soil can be thought of as fossils left over from ancient biomes. Some biomes, such as Temperate Forests, tend to have rich soils, while most other biomes have unproductive soils for agriculture. There are also important regional variations that make some areas unexpectedly productive or unproductive.

The US Department of Agriculture has developed a soil taxonomy that is very useful for classifying soils by their potential productivity for agriculture. To simplify their system, I have grouped soil types into three categories based upon their potential for agricultural production: Productive, Somewhat Productive and Unproductive. Each category denotes the productivity for agriculture, not plants in general.

A full-color version of this map can be found at my website. See the front matter of this book for a URL and a QR code to the website.

Alfisols (light green on the map) and **Inceptisols** (orange on the map) are both productive soil types for agriculture. Easily worked by humans using hoes or animal-driven plows, these two soil types have been a foundation for agricultural societies for centuries.

Notice that these soil types are far more common in Europe, East Asia, South Asia, Australia and New Zealand than in the rest of the world. Together they cover about 25 percent of ice-free land (10% for Alfisols and 15% for Inceptisols).

Soil Types

Legend

Productive for Agriculture:
- Alfisols
- Inceptisols
- Andisols

Somewhat Productive for Agriculture:
- Ultisols

Productive with Steel Plows (after 1830 only):
- Mollisols

Not Productive for Agriculture (white on map):

Andisols	Gelisols	Oxisols
Entisols	Histosols	Spodosols
Vertisols	Ice/Glacier	Shifting Sand
Rocky Land		

Source: US Dept. of Agriculture Natural Resources Conservation Service

Full color graphics available at:
frompovertytoprogress.com

Somewhat less productive, but still capable of supporting agricultural societies are **Ultisols** (yellow on the map). Because this soil type is much older than the previously mentioned soil types, its nutrients have tended to be leached over time. Despite this weakness, this red clay soil was the foundation of civilizations in Southern China, Southeast Asia, Korea, Japan and the Southern USA. Ultisols cover about 8% of ice-free land.

One very rare soil type that is a bit of a wild card in terms of its agricultural productivity is the volcanic **Andosols** (purple on the map). While many stretches of this soil type have not supported agriculture, sections of Andosols were critical to the civilizations in Mesoamerica, the Andes, Java and Japan. Andosols cover only about 1% of ice-free land.

The most productive soil for agriculture that covers large areas is **Mollisols** (dark green on the map), which covers seven percent of ice-free land. Mollisols are deep, rich soil with a high content of organic matter. This soil type is largely found beneath Temperate Grasslands in North America, Ukraine, Central Asia, the Fertile Crescent and Argentina. This incredibly rich soil is the foundation of many of the most agriculturally productive regions in the modern era.

The dense root mat that gives the soil its richness is, unfortunately, also the primary problem with exploiting this soil. Until the invention of the steel plow in the 1830s, humans could only break the sod with a huge amount of effort. This made dense human populations unlikely on this soil type until after the 1830s.

While the other seven soil types are theoretically capable of supporting agriculture in some circumstances, they have not done so in practice. This means that 61% of ice-free land is highly unlikely to ever support the dense human populations that are necessary to promote innovation and progress.

Growing Season

KEY INSIGHTS

Most regions have growing seasons that are too short for productive agriculture (the first key to progress). They are too hot, too cold, have too little precipitation or too much.

Even if a region has a desirable biome, low altitude and agriculturally-productive soil types, other factors can preclude the development of productive agriculture. Even in areas that have all the necessary factors, some days are too hot; some days are too cold; some days have too much rain and other days have too little rain. One or two days of any of the above are not a problem, but if those days are strung together for weeks or months, agricultural production is seriously constrained.

The International Center for Agricultural Research in the Dry Areas (ICARDA) has created a metric called Growing Season, which defines whether a region has optimal amounts of sunshine and water on any given day. For a day to be defined as part of the growing season it must not be too hot, too cold, nor can it have too much nor too little precipitation. Unfortunately, I have only been able to find solid data for Eurasia and Northern Africa.

Just as we have seen from other geographical factors, only a few regions have a long enough growing season to support intensive agriculture. Southern China is by far the biggest region with a growing season of over 330 days per year. Parts of Korea, Japan, Burma and India are also blessed with extremely long growing seasons. Neighboring regions in upper Southeast Asia have Growing Seasons over 210 days per year.

The second large region with a Growing Season of over 210 days per year is the bulk of Europe. While Europe's Growing Season is far shorter than Southern China's, it is far longer than the remainder of Eurasia.

Entire swaths of Northern Asia, Central Asia, Western Asia and North Africa have very short or non-existent Growing Seasons. Like many of the other factors that we have mentioned previously, short Growing Seasons place a fundamental constraint on the evolution of productive agriculture, the first Key to Progress.

Proximity to the Middle East

In addition to the purely geographical constraints, some constraints have been partly geographic and partly man-made. The most important such constraint was proximity to the Middle East. As we have seen, agriculture evolved in the Fertile Crescent (modern-day Syria, Iraq and Southern Turkey). Early Agrarian societies in Egypt, Mesopotamia and Anatolia played an important role early in the Agrarian era. These regions played a critical role because they had many wild ancestors of domesticated plants and animals.

KEY INSIGHTS

Being close to the Middle East enabled many regions to acquire plants and animals necessary for productive agriculture (the first key to progress). Those that were separated by oceans had no chance to do so.

As the peoples from those regions migrated westward to Europe, southward to Egypt and eastward to Central Asia and South Asia, they brought their domesticated plants and animals with them. Because annual rainfall patterns from the Indus River in modern-day Pakistan and the Mediterranean Europe are similar, crops could spread easily.

All of this meant that regions that lacked wild ancestors of domesticated plants and animals could receive them via migrants from the Middle East. This enabled regions that were relatively close to the Middle East to overcome the otherwise critical constraint of lacking the wild ancestors of domesticated plants and animals.

Proximity to Steppe Herding Societies

As Peter Turchin has argued, until the widespread use of firearms and cannons, the horse archers of the Eurasian steppe were the dominant military threat to Agrarian societies. Combining rapid strategic and tactical mobility with standoff weapons in the form of the composite bow-and-arrow, the Herding societies of the steppe were dangerous enemies.

Peasants in small villages did not stand a chance when even a small group of horse archers chose to attack. And given their propensity to destroy any vestiges of Agrarian society that they came across, one defeat could devastate a village for generations. Agrarian militaries, while very powerful, usually lacked the strategic mobility to intercept horse archers before they could attack.

KEY INSIGHTS

Being close to the warriors living on the Central Asian steppe made many Agrarian societies in Eurasia vulnerable to military conquest and plunder.

Agrarian societies that were located near Temperate Grassland biomes with horse archers faced an existential threat. The histories of East Asia, Central Asia, South Asia, Russia and the Middle East are full of military conquests by horse archers from the steppe.

The first phase of military conquest usually consisted of wholesale slaughter and devastated cities. Usually, the occupiers from Herding societies settled down and placed themselves in charge of the extractive institutions that Agrarian elites had so conveniently already put in place. Over time ethnic tensions between the Herding elite and the Agrarian masses led to the collapse of these regimes, but it often took Agrarian societies generations to recover from the original devastation.

As this perpetual cycle of conquest, devastation, exploitation, liberation and recovery continued, it had long-term negative effects on economic development. Agrarian elites were forced to extract additional resources from the already over-stretched peasantry to pay for defending the realm.

Sometimes defense spending by Agrarian regimes took the form of bribing nearby Herding societies to not attack. Other times it took the form of building huge defensive walls: the Great Wall of China being the most famous. Other times, it consisted of recruiting rival Herding societies into mercenary armies of horse archers. Whichever strategy Agrarian elites chose, it was never good for economic development.

As Peter Turchin has argued, Agrarian societies in East Asia, South Asia, the Middle East and Russia were dominated by military competition with the Horse archers of the Central Asia steppe. This forced those Agrarian societies to grow large and ruthlessly extract food surpluses to survive. This centralized extraction stifled innovation.

Agrarian societies of Western Europe and Southeast Asia faced much less of a military threat, so they evolved into more decentralized societies. As we have seen, the political decentralization of Northwestern Europe was a key factor in the rise of Commercial societies and progress.

Conclusion

Differences in contemporary per capita GDP are rooted in deep historical factors. The most important historical constraints on the development of complex human societies are geographical constraints to food production. To grow a food surplus capable of supporting large cities, a society must be in a region with very specific geographical characteristics.

Where geographical factors are conducive to agricultural systems based upon animal-driven plows, complex societies capable of sustaining innovation are likely to develop. Where those geographical factors placed serious constraints, a society had no chance to do so unless they came into contact with more complex societies.

As we will see, far more regions were unfortunately highly constrained. This had a long-term effect on the evolution of society types, innovation, progress and economic development. This created a trap that has persisted for millennia.

The key geographical factors that enabled or constrained food production and the growth of complex societies were:

- Suitable biomes
- Habitat for domesticable plants, particularly rice, wheat and corn.
- Habitat for domesticable animals, particularly horses.
- Rivers
- Altitude
- Soil type
- Growing season
- Proximity to the Middle East
- Proximity to Steppe Herding societies

In the next chapter, we will examine how these factors influenced each of the regions on planet Earth.

GEOGRAPHICAL CONSTRAINTS BY REGION

I n this chapter, I will examine each of the regions on planet Earth to see whether their local geography enabled or constrained the evolution of complex human societies. Without complex human societies, progress is not possible.

The most important characteristic of a geographical region is which society types can ultimately evolve within it. Since society types are based on how humans living within that society acquire enough food to survive and reproduce, the critical link between geography and society type is which types of food production a natural environment can potentially support. Humans then invent new food production and distribution technologies, skills and organization to maximize the potential within their natural environment.

I believe that given enough time, if a society type can evolve in a specific region, sooner or later a group of people will figure out how to do it.

Sequencing of Society Types

In any one natural environment, there are only a few society types that can potentially evolve within it. This leads to what I call the "sequencing of society types". Rather than a linear sequence of societies naturally moving from simpler to more complex societies, we see a series of branching sequences based upon natural environment and technology.

Because Hunter-Gatherer, Fishing, Herding, Horticultural and Agrarian societies require radically different geographical conditions, they lead to five different "dead-end" sequences. By "dead end", I mean that those societies cannot evolve The Five Keys to Progress within their own region.

In other words, in many different natural environments, a society cannot transition to a more complex society type. For this reason, any group of people who live in those regions are trapped in poverty by geographical constraints. This means that modern-day progress was impossible without Industrial technologies, skills and social organizations being introduced from other societies located in different natural environments.

The five different sequences that can potentially occur within a geographical region are:

- Hunter-Gatherer sequence: Hunter-Gatherer [stop]
- Fishing sequence: Hunter-Gatherer > Fishing [stop]
- Herding sequence: Hunter-Gatherer > Herding [stop]
- Horticultural sequence: Hunter-Gatherer > Horticultural [stop}
- Agrarian sequence: Hunter-Gatherer > Horticultural > Agrarian [stop, at least until 1830]

In the first sequence, geographical factors enabled the evolution up to, but not beyond, Hunter-Gatherer societies. In the second sequence, geographical factors enabled the evolution of Hunter-Gatherer societies and then Fishing societies, but no further. In the critical Agrarian sequence, geographical factors enabled the evolution of Hunter-Gatherer societies, then Horticultural societies and finally to Agrarian societies. Then critical geographical constraints created a cul-de-sac that precluded further development until recently.

In later chapters, we will explore the more complex sequences that result in Commercial and Industrial societies, but for now, these are five significant sequences. The sequence that we need to be on the lookout for while we survey each geographical region is the Agrarian sequence as it is the most complex pre-modern society. And as we will later see, it was a necessary intermediate step to Commercial and Industrial societies capable of delivering progress.

Now that we have identified the critical factors that led to the creation of complex Agrarian societies, let's take a tour of each region. Because I will focus on geographical factors, the regions that I will use differ somewhat from the regions based upon political boundaries that are more typically used. In general, I will use the border of biomes and continents, rather than political boundaries, to separate the world into regions.

Sub-Saharan Africa

The evolution of complex societies in Sub-Saharan Africa has been seriously constrained by geography. The biomes in Sub-Saharan Africa are not conducive to the evolution of complex societies. Africa is dominated by the Tropical Forest, Savanna and Desert biomes.

The biome that is most conducive to development — Temperate Forest — is completely missing from the continent. Only a very small Mediterranean ecoregion exists near the Cape in the extreme south. This ecoregion is too small to enable the evolution of complex societies.

In addition, the non-Forest ecoregions in Africa experience extremes in precipitation. Rain is almost entirely concentrated in the Rainy season, while the Dry season is close to Desert conditions.

Most soil within Sub-Saharan Africa is not very productive. The soils of the Congo, East Africa and the Kalahari Desert are highly unproductive. There are, however, many regions with reasonably good soil. Inceptisol and Ultisols were fairly widely scattered throughout the region, plus the broad swath of productive Alfisol soil in the Sahel and

Southeast Africa could have supported productive agriculture.

Another major disadvantage for Sub-Saharan Africa is that the bulk of the region rests on an enormous plateau that is over 3000 feet in altitude. The edge of this enormous plateau is usually very close to the ocean, leaving little room for coastal plains. At the same time, there are few large mountains that can support a Tropical Highlands biome, the most advantageous biome in Tropical latitudes.

> **KEY INSIGHTS**
>
> Sub-Saharan Africa had many geographical constraints that made it impossible to evolve into Agrarian societies. This made it impossible for them to acquire The Five Keys to Progress without outside help.

Rivers in Sub-Saharan Africa are hardly conducive to the evolution of complex societies. With few mountains tall enough to accumulate snowpack, little water is stored for the dry season. While the rainy season turns African rivers into torrents, the subsequent dry season turns those same rivers into a relative trickle. This makes navigation by large boats difficult, if not impossible, in both seasons. In addition, the proximity of the African plateau to the coastlines means that large rivers tend to have large waterfalls and cataracts near their mouth, another serious limitation to navigation between the ocean and the interior of the continent.

Sub-Saharan Africa did have some native wild ancestors of domesticable animals and plants. The Sahel region (just south of the Sahara desert) was a source of rice, millet and sorghum, giving the region a viable base of staple crops. East Africa is known for its large herds of herbivores, but unfortunately, none of them were suitable for domestication. Though Sub-Saharan Africa did adopt cows, pigs, and sheep from the Nile river valley, horses were not used until very recently. This placed a major constraint on African Herding societies.

Nor does Sub-Saharan Africa have coastlines that could have supported port cities. The African coastline has very few peninsulas or islands, no major inland seas or protected waterways. And much of the continent is distant from the ocean. The combination of unnavigable rivers, great distances from the oceans and few good locations for ocean ports erected huge geographical barriers to African development.

If that is not enough, Sub-Saharan Africa is an ideal climate for two scourges of humans and domesticated animals: malaria and sleeping sickness. Malaria was, and unfortunately still is, a mass killer of humanity. Even those that survive the deadly disease suffer recurring bouts that sap energy and make the victim unproductive.

In addition, sleeping sickness, carried by the tsetse fly, created vast regions where cattle and other herd animals could not live and other zones where they could only live seasonally. No other region on Earth has been so constrained by its disease environment.

The regions within Sub-Saharan Africa that could support Herding societies were also a double-edged sword. Those Herding societies led raids on neighboring Horticultural societies. These Herding societies often conquered entire Horticultural societies and established themselves as the de facto ruling class.

One example that is very relevant today is in Rwanda, where the cattle-herding minority Tutsi dominated the majority horticultural Hutus. Resentment about centuries of domination was a key ingredient in the genocide that some Hutus carried out against the Tutsi population in 1994.

Finally, even though Sub-Saharan Africa is close to Eurasia, it is still relatively isolated. The Sahara Desert to the north, the Red Sea to the northeast and the Indian Ocean to the east each make it difficult for African societies to encounter and copy technologies, skills and social organizations from more complex Eurasian societies. Alone, these barriers would not have been insurmountable, but in conjunction with the other factors listed above, they hindered development in Sub-Saharan Africa.

All of these factors make it very unlikely that any complex Agrarian societies could ever have developed in Sub-Saharan Africa. While the Tropical Forest biome supported Horticultural societies and the Savanna supported Herding societies, more complex societies could not evolve. While Sub-Saharan Africa could support the Hunter-Gatherer,

Fishing, Herding and Horticultural sequences, it could not support the critical Agrarian sequence.

The biggest exception to this rule was the very small South African region of the Mediterranean biome with productive Alfisol soil. Because it was a small region with very different rainfall patterns (wet winters, but dry the rest of the year) and it lacked native wild ancestors of domesticable plants or animals, local Africans could not develop Horticultural or Agrarian societies in the area.

Before the arrival of the Dutch, the area that is now South Africa was dominated by tiny Khoisan Hunter-Gatherer and Fishing societies. The arrival of European settlers, who brought with them Agrarian technology, domesticated plants and animals enabled an explosive growth of Agrarian societies in this region.

South America

While less constrained than Sub-Saharan Africa, geography also limited the evolution of complex societies in South America. Tropical Forest, Tropical Highlands, Desert and Grassland biomes dominate South America. These biomes could support Horticultural societies, but nothing more complex.

The biomes that were most conducive to complex societies — Temperate Forest and Mediterranean — were almost completely absent. There are only two very small ecoregions — both in Chile. But these regions were small and isolated enough that they were unlikely to have supported large populations.

Unlike Sub-Saharan Africa, however, the towering Andes mountain range supported a Tropical Highland ecoregion. The high altitude of this region created significantly cooler temperatures than in the hot steamy lowlands. In addition, the Altiplano, a gigantic plateau in Peru and Bolivia, created a natural trap for soils conducive to agriculture. As we will see, this ecoregion accounted for the only complex Horticultural societies on the continent.

South America did, however, have many areas with fertile soil. Present-day Argentina has a large area of Mollisol soil, the most productive soil type in the world. Unfortunately, this soil could not be exploited until the invention of the steel plow in the 19th Century. Parts of present-day Argentina, Paraguay, Bolivia and Eastern Brazil have productive Alfisol soil. The critical Andes region and the lower half of Chile have large amounts of productive Inceptisol soil.

The Andes had a range of native wild ancestors of plants that were suitable for domestication. Potatoes, quinoa, squash and beans gave the region a reasonably healthy variety of edible foods. When maize (also known as corn) from Mesoamerica reached the Andes, the region finally had a nucleus of staple crops capable of supporting substantial populations.

The entire continent, however, lacked domesticable animals, with the exception of guinea pigs, turkeys, llamas and alpacas. Only llamas and alpacas in the Andes had a substantial impact on agricultural societies in South America. While llamas and alpacas could be used as pack animals, they were not strong enough to pull plows. The lack of traction animals was to be the single largest constraint on South American development.

Even the Temperate grassland of the pampas in Argentina lacked Herding societies, as there were no large domesticable animals that could be herded until the Spanish introduced horses and cows. Before the arrival of the Spanish, the Temperate Grasslands were only suitable for Hunter-Gatherer societies.

South America does have many large rivers, such as the Amazon, Rio de la Plata and the Orinoco rivers, but none of them flow through biomes that are conducive to Agrarian societies. If they had flowed through Temperate Forest or Mediterranean biomes, these rivers might have become the backbone of complex Agrarian societies, but that did not happen.

Unlike in Africa, diseases were not a significant barrier to the evolution of complex societies in South America. Malaria was not native to the region. European settlers and their African slaves brought the disease to South America after 1500. Because South America had no wild ancestors of domesticated animals, there was no military threat from Herding societies.

Most importantly, the Pacific and Atlantic Oceans formed an insurmountable barrier before 1500, making it impossible for South American societies to copy innovations made in Eurasia. Even within the New World, South America was relatively isolated. The narrow Central American Isthmus and the Tropical Forest biome in that area isolated South America from North America and Mesoamerica.

All of these geographical factors meant that complex Agrarian societies could not evolve in South America until the arrival of Europeans. While South America could support the Hunter-Gatherer, Fishing and Horticultural sequences, it could support neither Herding nor the critical Agrarian sequence.

By far the most complex societies in South America were the Horticultural societies of the Andes Mountains. Far less populous and technologically complex Horticultural societies also evolved in the Tropical Forests of the Amazon and other regions. But the bulk of the continent was dominated by Hunter-Gatherer and Fishing societies.

The arrival of European settlers in present-day Argentina, Uruguay, Chile and Southern Brazil, areas with biomes and soil conducive to complex societies, led to the explosive growth of Agrarian societies in those areas. As the settlers brought complex Agrarian technologies, particularly domesticated plants and animals, the region's geographical advantages could suddenly be exploited.

Central America and the Caribbean

The regions of Central America and the Caribbean suffered from even greater constraints than their neighbors to the south. Virtually the entire region is covered by Tropical Forests, limiting development to Horticultural societies. While some mountainous regions could potentially have supported a Tropical Highlands biome, they were very small and lacked large plateaus analogous to the Altiplano in the Andes.

In addition, Central America and the Caribbean had no native wild ancestors of domesticable animals or plants. Nor did the regions have any large rivers. It is easy to see why the region never developed beyond simple Horticultural societies before the arrival of the Spanish.

Even after the arrival of the Spanish, Central America and the Caribbean remained backwaters within the Spanish empire (except for Cuba and Panama).

As we saw with South America, Central America and the Caribbean could support the Hunter-Gatherer, Fishing and Horticultural sequences, but they could not support the critical Agrarian sequence.

> **KEY INSIGHTS**
>
> The Americas had many geographical constraints that made it impossible to evolve into Agrarian societies. This made it impossible for them to acquire The Five Keys to Progress without outside help.

Mesoamerica

Mesoamerica as I am defining it covers Central Mexico plus the northern half of Central America. In terms of geographical factors, the region shares more with the Andes than to closer regions.

Mesoamerica is the homeland of many native wild ancestors of domesticable plants. By far the most important is maize (corn), one of the three most important staple crops in the world. The region also spawned the native wild ancestors of beans, squash, chile peppers, cotton and sweet potato, giving the area the most diverse agricultural base of any in the New World.

The second big advantage for Mesoamerica is its soil. The region is covered with productive Alfisol and Inceptisol soil, and the somewhat productive Ultisol soil. Even better, Central Mexico is blessed with some of the largest tracts of volcanic Andosol soil in the world. This soil formed the foundation for perhaps the most complex Horticultural society ever to evolve: the Aztecs.

Mesoamerica is dominated by a wide variety of Forest biomes: Tropical Rain Forest, Seasonal Forests and Tropical Coniferous Forest, the last of which is an extremely rare biome. Unlike most regions in tropical latitudes, Central and Southern Mexico has a large portion of land in high-altitude plateaus. While these areas do not technically qualify as a Tropical Highland biome, they do offer an excellent habitat for Horticultural societies.

The side effect of these highlands was that Mesoamerica had no navigable rivers. The region did have significant coastlines, but the locals rarely seemed to use them for long-distance navigation. This placed serious constraints on transportation within Mesoamerica.

The single biggest geographical disadvantage suffered by Mesoamerica is a lack of native wild ancestors of domesticable animals capable of pulling a plow. Lacking horses, cattle and water buffalo meant that agriculture based upon animal-driven iron plows was impossible. This placed a hard constraint on the possibilities for developing a large food surplus. The only advantage of this is that there was no military threat from Herding societies.

As we saw with South America, Mesoamerica could support the Hunter-Gatherer, Fishing and Horticultural sequences, but it could not support the critical Agrarian sequence. All of these geographical factors made it very likely that the region would see the rise of Horticultural societies, but that they could never make the jump to more complex Agrarian societies until the arrival of Europeans.

North America

No other region on the planet has such a wide variety of geographic advantages and constraints as North America. Like the rest of the New World, North America is isolated from Eurasia by the Atlantic and Pacific Oceans. In addition, the Sonoran Desert separates the region from Mesoamerica. The Sonoran Desert is not an impenetrable barrier, particularly along the coasts, but this arid region made it much more difficult for North American societies to acquire innovations from Mesoamerica.

North America has a wide variety of biomes. The Tundra and Boreal ecoregions in the north and the Desert and Temperate Coniferous ecoregions in the West are not conducive to the evolution of Agrarian societies. The vast Temperate Forests of present-day Eastern United States and parts of Canada, however, are especially conducive to complex societies. And once the steel plow was invented in the 1830s, the Temperate Grasslands of the Great Plains became the largest and most productive agricultural region in the world.

North America also has many regions with soil that are suitable for agriculture. Mollisol soil, the most productive soil in the world, dominates the Great Plains. Only the Asian steppe rivals it in size. Productive Alfisol soil is common in the Upper Mississippi, Ohio and Sacramento River basins, and the northeastern United States has large areas of productive Inceptisol soil. Though not as productive as the other soils mentioned above, Ultisols make the Southeastern United States agriculturally productive.

The bulk of North America is also well below 500 meters in altitude. The western-third of the present-day United States and the Appalachian mountain chain are the only large regions where the altitude is a major constraint to the evolution of complex societies.

North America is also blessed with both the Mississippi-Missouri-Ohio river basin, and the St. Lawrence-Great Lakes waterways. These two enormous bodies of water form liquid superhighways for societies with advanced boating technologies. The only portion of the Eastern United States that does not have large navigable rivers is the Atlantic seaboard with its many natural ocean harbors, such as Boston and New York City.

The principal barrier to the evolution of complex North American societies was a big one: a lack of native wild ancestors of domesticable plants and animals. There were some native plant species, such as sunflower, squash, sumpweed, maygrass, goosefoot and little barley, but staple crops capable of supporting large populations were missing. It was not until maize diffused north from Mesoamerica into the present-day Eastern United States that the region could potentially support large populations.

An even bigger barrier for North America was the lack of domesticable animals, except for the turkey. Ironically, horses had probably

> **KEY INSIGHTS**
>
> North America had almost all the necessary geographical factors to evolve into Agrarian societies, but the region was missing domesticatable animals capable of pulling plows. Once they acquired these animals from Europeans, the region grew spectacularly.

evolved in North America before they spread to Central Asia, but all prehistoric horse species had died off long before agriculture evolved. Whoever (or whatever) killed the last wild horse in North America dealt a death blow to the possibilities of developing Agrarian societies in that region.

The sudden emergence of Herding societies on the Great Plains after the Spanish introduced horses shows how quickly a society can change when key missing factors are introduced to a favorable environment. Before the arrival of horses, Native Americans largely avoided the Great Plains except along the banks of rivers. Once wild horses arrived, Native Americans very quickly mastered all the basic skills and technologies necessary to form Herding societies. A similar transition occurred in the Pampas of Argentina.

These emerging Herding societies quickly came into contact with each other and waged war for supremacy of the Plains. From this violent competition, the Comanche Nation formed a powerful empire that closely resembled those of the Eurasian steppe herders. This strongly suggests that North American history in the pre-conquest era would have been very different if it had possessed wild horses.

As we saw with the rest of the Americas, early North America could support the Hunter-Gatherer, Fishing and Horticultural sequences, but it could not support the critical Agrarian sequence. Despite having many geographical factors that were conducive to the evolution of Agrarian societies, the lack of domesticable animals capable of pulling plows placed a hard ceiling on their level of development. While the Eastern and Southwestern United States developed many simple Horticultural societies, most indigenous people lived in either Hunter-Gatherer or Fishing societies.

The arrival of European settlers, particularly the British, unlocked the enormous potential for development in North America. The British settlers brought with them technologies, skills and social organizations from complex European societies. Just as important, they brought with them wheat, cattle and horses.

In doing so, British settlers eliminated the final geographical constraints to the evolution of complex Agrarian and Commercial societies. The result was one of the most rapid periods of economic growth in world history. Within just a few generations, North America was transformed from a sparsely populated region with a relatively low technology base and standard of living to a densely populated region with a substantial technology base and a high standard of living.

Northern Eurasia

Northern Eurasia, as I have defined it, includes the Boreal and Tundra biomes in Siberia and northern Scandinavia. One does not have to be a geographer to know that this region is hostile to the evolution of complex Agrarian societies. With its long, dark winters, extreme cold and driving winds, Northern Eurasia makes life a contest for survival.

Northern Eurasia does have some environmental advantages. Virtually the entire region is below 500 meters in altitude, and large rivers cut through most portions of the region. Surprisingly, large tracts of Central and Southeastern Siberia are covered with productive Inceptisol soils, though no society has found a way to exploit this natural resource due to the short growing season.

In addition, Northern Eurasia is not as isolated from the complex societies of Eurasia as most other regions on the planet. The Boreal forests of Eastern Siberia come into direct contact with the northern reaches of the North China Plain, while the western edge comes into contact with the Temperate Forests of European Russia. And the region's southern border hugs up against the Central Asian steppe.

Northern Eurasia, therefore, had plenty of opportunities to come into contact with and then copy the civilizations of East Asia and Europe. Northern Eurasia was also vulnerable to invasion from steppe Herders, but since the region had little wealth worth plundering, it was not a problem in practice.

The vast river systems of Northern Eurasia might be expected to enable a dense trade network with other complex societies in Eurasia, but these rivers are not as advantageous as they first appear. Most rivers in Northern Eurasia are iced over for the bulk of the year, and almost all of them flow into the ice-bound Arctic Ocean, so they do not offer a practical route to the ocean. Any ship that traveled to the mouth would find that there were few towns worth trading with locally, and it would require a long dangerous path to reach the wealthy cities in Europe or East Asia.

The flatness of the terrain, usually an advantage, is turned into a disadvantage by the northward orientation of the rivers. The spring thaw comes to the warmer southern regions first, so the waters cannot flow northward through the still-frozen rivers. Nor can the icemelt percolate down through the still-frozen soil.

So the icemelt backs up and forms a labyrinth of lakes, ponds, puddles and waterlogged soil. It is difficult for most plants to grow on this waterlogged terrain, particularly those that are fit for human consumption.

Northern Eurasia also lacks native wild ancestors of domesticable plants and animals. Local Herding societies did manage to domesticate yaks and reindeer, but these species never reached the importance of horses, cows and other domesticated animals. Most importantly, neither can be trained to pull a plow.

Northern Eurasia could support the Hunter-Gatherer, Fishing and Herding sequences, but it could support neither the Horticultural nor the critical Agrarian sequence. For this reason, Northern Eurasia has had very little impact on human history. The regions gave birth to only Hunter-Gatherer, Fishing and Herding societies, all of whom had small populations. It was only with settlement by Agrarian societies, in this case, Russians and Scandinavians, that the region was able to transition somewhat beyond those states.

Southeast Asia

My definition of Southeast Asia is a little different from most. I include all the Tropical Rain Forests and Seasonal Forest ecoregions stretching from present-day Indonesia, northward to present-day Southern China and then westward to most of present-day India.

Southeast Asia's two biggest geographical advantages are 1) proximity to China, and 2) a dense network of rivers originating in the Himalayas and Tibetan plateau. When the Indian subcontinent collided with the Asian continent about 50 million years ago, the Himalayas grew up until they became the tallest mountains in the world. This towering mountain chain greatly impacts local wind patterns and led to the monsoon weather cycle.

KEY INSIGHTS

Southeast Asia had almost all the necessary geographical factors to evolve into Agrarian societies, and they were able to acquire domesticatable plants and animals from China and India.

Monsoon rains from Southeast Asia in the summer dump huge quantities of water into the foothills of the Himalayas. Since the Himalayas are comparatively young, they are steep and easily erodible. The combination of torrential rains and erosion created a vast network of rivers carrying massive amounts of silt into the valleys of Southeast Asia.

Southeast Asia has great variations in altitude. Much of the region is mountainous, although these mountains rarely are tall enough to make them uninhabitable. Rivers cutting through those mountains form steep gorges, at least until the rivers reach the sizable coastal plains. These coastal plains slow down the river flow enough so that they dump their silt in the river delta, a perfect recipe for fertile soils.

If Southeast Asia had been isolated from the rest of the world, people living next to these relatively small river deltas would probably never have developed the subsistence technology to support Agrarian societies. Fortunately, these regions were a short sail from both China and India, meaning that Southeast Asia could easily copy the skills, technology and organizations from those regions. The most important of the imported innovations were the plow, water buffalo and rice.

Geography and technology combined to create a settlement pattern unique to Southeast Asia: densely populated Agrarian societies in scattered river deltas, surrounded by vast highlands inhabited by sparsely populated Horticultural societies. These highland Horticultural societies were often of different ethnicities, religions and languages from the people living in the river deltas.

Among the most important river deltas of Southeast Asia were (from west-to-east):

- The Ganges-Brahmaputra-Megha river basins in present-day India and Bangladesh.
- The Irrawaddy River delta in present-day Burma.
- The Chao Phraya River delta in present-day Thailand.
- The Mekong River basin in present-day Cambodia and Vietnam.
- The Red River delta in present-day Vietnam.
- The Pearl River delta in present-day Southern China.

Even today, the populations of Southeast Asia are heavily concentrated in these small river deltas. Where the rivers run through relatively sizable plains before they run into the ocean, for example, in Cambodia and Southern Thailand, sizeable populations away from the river can evolve. And where the rivers run through extended plains less than 500 meters in altitude, such as in India and Bangladesh, huge populations developed.

The third, and historically least important region in Southeast Asia is the dense Tropical lowland jungles of present-day Indonesia, Malaysia and the Philippines. Societies in these regions were hampered by low altitudes, the lack of large rivers and malaria, so they could only support limited populations.

Southeast Asia also had a wealth of native wild ancestors of domesticable plants, but none can be considered a staple crop. Most likely, rice arrived from China, giving the region a staple crop that they would need to form complex Agrarian societies.

The lack of native wild ancestors of domesticable animals potentially presented a serious problem, but proximity to China and India meant that, over the long run, domesticated animals could gradually be imported for use in the fields. Water buffalo, the dominant traction animal of Southeast Asia, probably arrived from China. Without proximity to China and India, it is unlikely that Agrarian societies could have evolved in Southeast Asia.

Southeast Asia had one great advantage over the other regions on the Asian continent: no real military threat from steppe Herding societies. While East Asia, South Asia, Central Asia and the Middle East were constantly being conquered by steppe Herders, Southeast Asia was generally untouched. The Mongol invasion of Burma and their failed invasion attempts of Vietnam and Java were the only real exceptions.

Virtually all of Southeast Asia is covered by the somewhat productive Ultisol soils, giving areas that were not near rivers a chance to develop agriculture. There were also scattered areas of productive Inceptisol soil throughout the region, plus large tracts in the Ganges river basin. This was another factor in the huge population density in that area.

There is one region in Southeast Asia that stands out from all the others: the island of Java in present-day Indonesia. Without rivers and covered by Tropical Rain Forests, Java does not appear to be any more conducive to large human populations than the rest of Indonesia and Malaysia.

Java, however, is blessed with large tracts of volcanic Andosol soil, which in some circumstances can be the most productive soil in the world. Andosol soil, plus the importation of rice and water buffalo, gave rise to Agrarian societies on the island. Java was probably the most geographically isolated of all Agrarian societies.

Southeast Asia is the first region that we have examined that was capable of supporting Agrarian societies. The region could support the Hunter-Gatherer, Fishing and Horticultural sequences as well as the critical Agrarian sequence. But it is important to note that the sub-regions within Southeast Asia that could support Agrarian society were quite limited in scope. By geographical size, most parts of Southeast Asia evolved along the Horticultural sequence.

The fact that complex societies in Southeast Asia could only evolve in isolated river deltas had a large cultural impact. The scattered nature of Agrarian societies in Southeast Asia is unique in world history. The Agrarian societies of East Asia, the Middle East and Europe were all grouped in close proximity. This allowed them to copy innovation from each other, but it also left them vulnerable to military conquest. The history of warfare and empire in the rest of Eurasia gave each of them a much tighter cultural and technological community.

The Agrarian societies of Southeast Asia, except for the one in the Ganges River valley in India, are separated from each other by hundreds of miles of rainforest, mountains and oceans. This gave the region its characteristic cultural and technological diversity. While this isolation made it harder to copy innovation from their neighbors, it also meant that they were far less vulnerable to military conquest.

Interestingly, none of the Agrarian societies of Southeast Asia appear to have built a large navy that could have been used for either defense or conquest. With their decentralized political and economic structure and extremely long, indented coastline and large navigable rivers, one might think that Southeast Asia is a good candidate for the evolution of Commercial societies (which we will examine in greater detail in the next chapter).

Whether Commercial societies did, in fact, evolve in Southeast Asia is unclear. It is clear that there was a great deal of nautical trade in the waterways between India and China. This trade also led to the formation of trade-based cities, particularly in the critical navigational choke point of the Strait of Malacca between present-day Malaysia and Indonesia.

But did these cities qualify as full-fledged Commercial cities? They were clearly engaged in trade, plus they also had political autonomy. But their population sizes were small, and individual towns seem to have gone through rapid booms and busts. From the historical record, it appears that the towns in the Straits of Malacca had very small permanent populations before contact with Europeans.

Currently, the historical record is too limited to make a full determination, but it is clear that these towns never came close to being populous Commercial cities similar to Venice, Amsterdam or London. And outside of the Straits of Malacca, there were no other towns that could potentially make the cut. Unless more evidence emerges, I tend to believe that these towns, however intriguing, did not qualify as Commercial societies.

Central Asia

My definition of Central Asia is a little broader than most, as it includes the Desert regions of present-day Iran, Afghanistan, Pakistan, western China and western India, plus the small portions of steppe that are technically on the European continent. Central Asia is a region full of Temperate Grasslands (the Steppe), Deserts, Temperate Highlands and scattered Coniferous Forests, typically forming long, narrow east-west bands across the continent.

Central Asia is perhaps unique in human history because it has been dominated by Herding societies for much of its history. Agrarian societies were rare and isolated in a few scattered river valleys. While Herding societies in other regions had little impact on human history, the Herders of Central Asia had a sizeable influence.

Central Asia has a stark contrast in soil types. The Desert and Mountain regions are almost devoid of fertile soil, except on the banks of its few rivers. The Steppes, however, cover a huge region of productive Mollisol soil. Before the invention of the steel plow in the 19th Century, the deep, dense mats of grass roots made this soil very difficult to plow. Even today, much of it remains unexploited by agriculture.

The southern portion of Central Asia is on a plateau, well above 500 meters in altitude. The Steppe areas and the northern Deserts areas are all below 500 meters in altitude. Because virtually all of Central Asia is remote from the ocean, and rivers cannot flow over the mountains to the south, much of the region is in the largest Endorheic basin in the world. Endorheic basins are regions where rainfall eventually flows into lakes or inland seas like the Caspian and Aral, rather than to the ocean as most rivers do.

The biggest advantage that Central Asia possesses is its proximity to China, Eastern Europe and parts of the Middle East. This meant that inhabitants could copy skills, technologies and social organizations from numerous Agrarian societies in neighboring regions.

It also meant that Central Asia was a veritable superhighway between the most important regions in Eurasia. The Silk Road was often the only means of contact between these regions for hundreds of years, giving the richer Agrarian societies reason to move their technologies into close contact with Central Asian societies.

The other big advantage for Central Asia is that most large wild mammals that were suitable for domestication are either native to the Steppe or flourish there. Horses, cattle, pigs, sheep and goats are all native to the region. This enabled Herding societies to expand to far greater population sizes than Herding societies in any other region.

The invention of wheeled carts, composite bows, the domestication of the horse, and learning the skill to ride them revolutionized Steppe society. Regions that had previously had little impact on other societies evolved into feared warriors who conquered vast empires.

While Herding societies in Africa were able to conquer agricultural societies on a small scale, the Herding societies of Central Asia were able to do so on a continental scale. Central Asian Herders formed a permanent security threat that shaped the military history of Asian Agrarian societies, particularly China.

Central Asia is the first region that we have covered where the Herding sequence predominated. While there were scattered oases where the Hunter-Gatherer, Fishing and Agrarian societies coexisted, Herding societies dominated this region.

New Guinea

New Guinea is not considered to be a separate region by most geographers, but since the island is one of the few regions that invented an indigenous form of agriculture, it is worth mentioning. New Guinea is highly isolated from the societies of the Eurasian continent. The Indonesian archipelago potentially offered a pathway from the Asian mainland, but in practice, there was little if any contact with the mainland. This made it unlikely that New Guinea could copy technologies, skills or social organizations from Eurasia. The only advantage was that there was no local military threat from Herding societies.

Except for a small section of Savanna immediately adjacent to Australia, New Guinea is covered by Tropical Rain Forests. Unlike most

Tropical Rain Forests, however, much of New Guinea is higher than 500 meters in altitude. The higher elevation meant cooler temperatures and far less risk of malaria, but also no navigable rivers. Because of these factors, the bulk of the population was concentrated in these Highlands.

New Guinea also possesses relatively fertile soil, unusual for a Tropical region. Much of the island is covered with productive Inceptisol soil, while most of the rest has very productive Andosol soil or somewhat productive Ultisol soil.

KEY INSIGHTS

Central Asia, New Guinea and Australia had many geographical constraints that made it impossible to evolve into Agrarian societies. This made it impossible for them to acquire The Five Keys to Progress without outside help.

Because of its isolation, it should not be surprising that New Guinea had no native wild ancestors of domesticable animals. The island did, however, have a wide variety of native wild ancestors of domesticable plants: taro, sweet potato, bananas, sago and yams. Noticeably lacking, however, were grains.

The combination of high elevation in the Tropical latitudes, numerous native wild ancestors of domesticable plants and productive soil meant that New Guinea was able to invent an indigenous Horticultural society. Its geographical isolation and lack of native wild ancestors of domesticable animals, however, made it impossible for those societies to evolve into Agrarian societies before the arrival of Europeans.

While the New Guinea lowlands were dominated by Hunter-Gatherer and Fishing societies, societies inhabiting the New Guinea highlands were able to transition along the Horticultural sequence. Fundamental geographical constraints made the Agrarian sequence impossible. Even after the arrival of Europeans, this has not changed.

Australia

The continent of Australia is even more isolated from the Eurasian mainland than New Guinea. Australia is unusual in that it included both Tropical and Temperate latitudes. Despite having a relatively small area compared to other continents, Australia has five different biomes: Savanna, Temperate Grassland, Desert, Temperate Forest, and Mediterranean.

The vast majority of Australia has unproductive soils, although the entire east coast and southwest corner are covered by productive Alfisol soil.

Australia is an example of a region of extreme geographical contrasts. The east coast and southwest have ideal biomes and soil for the development of complex Agrarian societies. But the combination of geographical isolation and a lack of domesticable plants or animals made it impossible for indigenous societies to evolve beyond Hunter-Gatherer or Fishing societies.

Australia was capable of supporting the Hunter-Gatherer and Fishing sequences. None of the other sequences were possible for this very isolated region. As with North America, Australian geography made it highly likely that, once Europeans migrated from more complex societies, these settler societies would expand rapidly.

The Middle East and North Africa

The Middle East as I am defining it includes Turkey, Egypt and North Africa, but excludes the bulk of Iran. The region is separated into two distinct sub-regions, the Desert biomes of Arabia and the Sahara with a jumble of Temperate Forest, Mediterranean and Temperate Grassland biomes to the north. The latter region became the homeland of the most important food production complex in history.

Soil also bifurcates the region: the Desert areas have very unproductive soil for agricultural purposes, while the North has large patches of Alfisol, Mollisol, Inceptisol and Ultisols.

The altitude profile of the region is almost the opposite. The Desert regions are at a desirable altitude of 500 meters or less, with the northern Highlands being over 500 meters high. The small "sweet spot" with lower altitudes and biomes and soil conducive to agriculture is called the Fertile Crescent, where agriculture first evolved.

The single largest advantage of the Middle East is that the region possessed more native wild ancestors of domesticable plants and animals than any other region. The Fertile Crescent had the native wild ancestors of wheat, barley, peas, beans, lentils plus cows, sheep and pigs.

The Middle East also possessed three large navigable rivers: the Tigris, Euphrates and Nile river valleys. These river basins provided the habitat for the Agrarian empires that would dominate the region for thousands of years. The Middle East also possessed long coastlines with many natural ports on the Mediterranean, Black Sea, Red Sea and the Persian Gulf. This made it relatively easy for the technologies, skills and social organizations that evolved in the region to diffuse into neighboring regions.

The Middle East is the first region that we have examined so far that possessed large geographical areas capable of following the Agrarian sequence. The region was the first to do so. Being in a central location within the Eurasian continent, the Middle East played a critical role in dispersing domesticated plants and animals to many other regions.

In this way, the Middle East played a major role in creating the progress that would later be enjoyed by other regions. It also made the region highly vulnerable to military threats from Herding societies.

East Asia

My definition of East Asia includes Japan, Korea and the North China Plain, but it excludes the Deserts and Highlands of western China and the Tropical Forests of Southern China. Other than the Middle East, East Asia developed agriculture and Agrarian societies earlier than any other region. Unlike the Middle East and other regions, East Asia has been dominated by one society: China.

The principal geographical advantage of East Asia is a huge expanse of Temperate Forest biome covering most of China and all of Korea and Japan. This gave the region vast tracts of land that could potentially be used for agriculture.

The North China Plain also had extensive tracts of Inceptisol soil.

Korea and Japan were less fortunate with their soil, although Korea has a section of Ultisol soil near present-day Seoul. And Japan had sections of Inceptisols and Andosols, particularly in Southern Honshu and the Osaka/Tokyo region.

Altitude had a similar profile as soil in the region. The North China Plain is less than 500 meters in altitude, except for the northwest section near Mongolia. Mountains over 500 meters high dominate the Korean peninsula, although the western coast has sizeable plains. Japan is even more dominated by mountains, but there is a small plain in the Osaka/Tokyo region.

While East Asia is not as blessed with large numbers of native wild ancestors of domesticable plants and animals as the Middle East, it did have the advantage of reasonable proximity. East Asia was able to supplement its biological resources with species that spread from the Middle East via Central Asia.

East Asia was also blessed with access to rivers and the ocean. The Yellow and the Yangtze are two of the largest navigable rivers in the world running through a Temperate Forest biome. In addition, the most heavily populated regions in China, Korea and Japan were close to the ocean.

For all of the reasons listed above, it should not be a surprise that East Asia saw the evolution of complex Agrarian societies throughout the region. East Asia possessed one of the largest geographical areas capable of supporting the Agrarian sequence. In particular, the North China Plain has played a critical role in world history. Not surprisingly Hunter-Gatherer and Fishing societies were pushed aside by the far more powerful Agrarian societies.

The main disadvantage that East Asia had was that it was extremely vulnerable to military invasion by Herders from the Central Asian steppes. Chinese history is littered with dynasties established by

> **KEY INSIGHTS**
>
> East Asia, Mediterranean Europe and Northern Europe had almost all the necessary geographical factors to evolve into Agrarian societies, and they were able to acquire domesticatable plants and animals from the Middle East.

Central Asians. The Xianbei, Di, Xiongnu, Jie, Qiang, Dingling, Sogdian, Gokturk, Shatuo, Khitan, Baiman, Tangut, Jurchen, Mongol, and Manchu are all examples of non-Chinese people who ruled all or parts of China for a significant period. The vast majority of these people lived in Herding societies from the steppe.

Mediterranean Europe

Mediterranean Europe includes Greece, Italy, Spain, Portugal and Southern France. Not surprisingly, given its name, Mediterranean Europe is dominated by the Mediterranean biome, which was capable of supporting complex Agrarian societies.

Mediterranean Europe had no native wild ancestors of domesticable plants and animals, but the region's proximity to the Middle East and similarities to its climate more than made up for this disadvantage. The region is extremely accessible from the Middle East, enabling inhabitants to easily copy innovations that enabled plow-driven agriculture.

The shores of the Eastern Mediterranean are effectively transition regions between the Fertile Crescent and the rest of the Mediterranean. Plants and animals that were domesticated in the Fertile Crescent made an easier transition to Mediterranean Europe than any other region.

Except for Greece, Mediterranean Europe has large tracts of Inceptisol soil. While there were very few large navigable rivers, the entire region was blessed with a long, jagged coastline with many potential port locations. Given that the Mediterranean was sheltered from strong winds and tides, and sailors only needed to traverse short distances between land, the sea made up a convenient "kiddy pool" for the development of maritime technology. Not surprisingly, virtually all recorded naval battles before the year 1500 took place in the Mediterranean Sea.

Mediterranean societies were also relatively free from military threats from Herding societies. The principal military threat to Agrarian societies in this region was other Agrarian societies.

The biggest disadvantage of Mediterranean Europe is that mountains over 500 meters in altitude cover the bulk of the region. This meant that most of the agricultural regions hugged the relatively narrow coastal plains, which were also vulnerable to erosion.

Mediterranean Europe had all the geographical conditions necessary to evolve along the Agrarian sequence. Primarily because of the Mediterranean biome, relatively productive soil, navigable bodies of waters and proximity to the Middle East, Mediterranean Europe evolved complex Agrarian societies. The most famous of these were Ancient Greece, Rome and Carthage.

The proximity to the first region that evolved along the Agrarian sequence — the Middle East — meant that Mediterranean Europe played an outsized role in ancient history. A region that was even more conducive to the evolution of complex Agrarian societies, however, would later eclipse it.

Northern Europe

In my definition of Northern Europe, I am including the entire European continent, except for those regions bordering the Mediterranean and northern Scandinavia (which were discussed earlier). Northern Europe has many geographical advantages that in combination are shared by no other region in the world.

Northern Europe has by far the largest expanse of the Temperate Forest biome in the world, running from the Atlantic coast eastward to well past the Ural Mountains in Russia. In addition, while Northern Europe is relatively distant from the Middle East, Mediterranean Europe and the Balkans offered transition regions between the two. This made it relatively easy for Northern Europe to adopt the skills, technologies and social organizations of the Middle East, even though they did so at a much later date than most other complex Eurasian societies.

Northern Europe also has large tracts of Inceptisol and Alfisol soils. The southeastern section has the largest tracts of Mollisol soil in the world. While the latter was not useful until the invention of the steel plow in the 19th Century, the region offered great potential. And virtually all of Northern Europe, apart from Norway and the Alps, has an altitude of fewer than 500 meters.

While Northern Europe had few native wild ancestors of domesticable plants and animals, the domesticated versions spread from the Middle East via the Danube River basin and the Mediterranean. This meant that the original lack of domesticable plants and animals did not serve as a constraining factor in the development of Northern European societies.

Northern Europe possessed a vast network of navigable rivers that flowed year-round and emptied into the ocean. The Seine, Meuse, Thames and Rhine rivers have played particularly important roles in the development of Northern European societies. The region also possessed a long, jagged coastline with many natural ocean ports. This enabled the evolution of port cities that drastically lowered the cost of transportation.

> **KEY INSIGHTS**
>
> These favorable geographical factors enabled numerous Agrarian societies to evolve along a dense belt that stretched across Eurasia. Nowhere else in the world had the geographical conditions necessary to evolve into Agrarian societies, leaving them far poorer.

The military threat from Herding societies of Eurasian grasslands varied greatly within this region. Eastern Europe was highly vulnerable until they adopted firearms and cannons. Modern-day Russia, Ukraine, Rumania and Hungary were situated on or close to the Temperate Grassland biome. In addition, the Ottoman Empire, originally a Herding people, dominated the Balkans for centuries. This vulnerability to the military threat from Herding societies is one of the main reasons why Eastern Europe developed more slowly than Western Europe.

Western Europe, however, was fairly insulated from invasion from Herding societies. While the Huns and Mongols formed existential threats during certain periods, the great distance from the Temperate Grasslands biome in Central Asia made it difficult for Herding societies to find feed for their horses. Eastern Europe effectively created a defensive wall that protected Western Europe from attack. The main military threat to Agrarian societies in Western Europe was largely from other Agrarian societies.

All of these geographical factors combined to make Northern

Europe a perfect environment for complex Agrarian societies.

Perhaps even more important for European development, the region provided a perfect geographical and political environment for the development of Commercial societies. The fact that it was the only region with a large cluster of Commercial societies was essential in Northern Europe's development into the richest region in the world.

Conclusion

The inequalities between peoples in the 20th and 21st Century can be explained largely by geographic constraints on how food could historically be produced. These constraints led to the evolution of very different types of societies to evolve.

Most regions had geographic constraints that forced them along an evolutionary sequence that made Agrarian societies impossible. This in turn made Commercial and Industrial societies impossible to evolve (at least until very recently). The most common explanations of contemporary regional inequalities — institutions, political leaders, government policy, trade, colonialism or racism — are merely symptoms of a much deeper problem.

Geography has played a powerful role in facilitating and constraining the evolution of complex societies. Even if we exclude the hostile regions of the ocean, polar ice caps and mountaintops, most regions have geographical characteristics that make complex Agrarian and Commercial societies unlikely.

Migrations played an important role in smoothing out regional differences within Eurasia. Skills, technologies and social organizations could first evolve in areas that were highly conducive to their discovery. Humans could then migrate to neighboring regions that were somewhat less supportive. Migrants would modify those skills, technologies and social organizations to be more effective in their new environment. This in turn created new variations that influenced other regions.

In some cases, as in Northwest Europe and North America, societies mastered skills, technologies and social organizations pioneered in other regions, but then added on another layer of complexity that could not have existed in the first region. European migrations after 1500, in particular, played a key role in spreading complex skills, technologies and social organizations throughout the world.

But outside of Eurasia, regions could not support food production systems based upon animal-driven plows. Worse, these societies were isolated from each other. Lacking the prerequisites for the development of complex societies as well as being out of contact with societies that did have those prerequisites guaranteed far slower development than in Eurasia.

North America, South America, Sub-Saharan Africa, Siberia, New Guinea and Australia had no chance of developing into Agrarian societies before contact with Europeans. The Agrarian sequence was just not possible in these regions. These societies lived in a poverty trap created by fundamental geographical constraints.

Hunter-Gatherer, Fishing, Horticultural and Herding societies cannot possibly generate the food surplus to enable the growth of trade-based cities that in turn enabled innovation and progress. Given these geographical factors that long predated the current era, there is no need to discuss the contemporary impact of institutions, political leaders, government policy, trade, racism, colonialism and exploitation.

The vast majority of human societies could not possibly have generated anything like the progress that we take for granted today without outside contact. Given this fact, it is somewhat of a miracle that progress exists at all.

THE BEGINNING
OF PROGRESS

S o far we have seen that most societies were incapable of evolving into complex Agrarian societies because of fundamental geographical constraints. Only a relatively small portion of the globe stretching from Western Europe, through the Middle East and South Asia to East Asia could do so. All other societies were incapable of fulfilling the first Key to Progress: a highly efficient food production and distribution system. Only Agrarian societies were capable of doing so.

We also learned that Agrarian societies rarely develop the next two Keys to Progress: trade-based cities and decentralized political, economic and religious power. In addition, elites in Agrarian societies created centralized extractive institutions that siphoned off the food surplus in their favor. This centralized power and stifled the growth of trade-based cities.

All of this meant that progress failed to take place across virtually the entire globe. With all of these geographical and political constraints, it seems likely that true, sustained progress could not happen at all.

So how did we get here? How did we get from a time when virtually no one experienced progress to the current day, when the majority of mankind is experiencing progress? How did we get from a time when no societies had the Five Keys to Progress to a time when most societies do?

The answer is that a few very unusual societies were able to overcome geographical constraints, escape the poverty trap and generate progress for their people. These societies evolved each of the Five Keys to Progress. When each of the Five Keys to Progress is in place, a society becomes a vast decentralized problem-solving network that promotes progress.

Human history can be seen as the gradual, unconscious building of these five keys. No one deliberately set out to do so, but the benefits of each of the five keys were great enough that humans living in specific types of societies were highly likely to eventually stumble upon them and adapt them to solve local problems.

Those solutions worked so well that other people would inevitably copy them, so the keys could spread throughout society. Finally, other societies could see the benefits of those solutions, copy them and bring the keys to their own society.

Transforming Food Surplus into Progress

Earlier we examined the process by which humans combined energy with technologies, skills and social organizations to create complex Agrarian societies. Continuing this process, a few societies learned how to transform Agrarian societies into a vast problem-solving network that generated real progress for the masses.

This process is the following:

1. In societies without centralized extractive institutions, the food surplus generated by increased agricultural productivity enabled farmers to trade with individuals who possessed skills related to non-agricultural technologies (for example bakers, blacksmiths, millers and butchers).
2. When trade reached a certain level, large trade-based cities with a wide variety of skills evolved.
3. These trade-based cities created a virtuous cycle of

technological innovation, skills acquisition and social cooperation. This virtuous cycle included:

    a. Increased population sizes, which led to more people
       who could potentially innovate.

    b. Increased urbanization rates, which put those people
       in close contact with each other.

    c. A greater number of market transactions.

    d. Acquisition of new skills to conceive of, design, build,
       test, use and repair new technologies.

    e. Greater specialization of skills.

    f. Greater cooperation between people within the
       same group.

    g. Expansion of the sphere of cooperation beyond just
       the family to create larger and more specialized social
       organizations.

    h. Higher levels of trust within the group.

    i. Lower levels of violence within the group.

    j. Increased positive-sum thinking (i.e. we can both get
       ahead by cooperating, rather than trying to sabotage
       each other by lying, stealing or killing)

    k. Greater appreciation of achievement over status and
       military glory.

    l. Greater willingness to copy other societies.

4. This virtuous cycle within trade-based cities also has a powerful
effect on institutions by:

    a. Deepening the complexity of existing institutions as
       they put new technologies and skills to use.

    b. Enabling the creation of new institutions based upon
       new technologies and skills.

    c. Allowing existing institutions that are unable to make
       effective use of new technologies to collapse from lack
       of revenue.

5. All of the above fed back into the technology base, enabling
a faster rate of technological innovation. Each new piece of

technology could then be used as a component in another more complex piece of technology.

6. Finally, the widespread usage of fossil fuels broadened the geographic scope of this process so that it encompassed more than just a few trade-based cities. The feedback loop could now affect entire regions, nations and eventually continents.

Breakthroughs that Promoted Progress

Six historical breakthroughs enabled progress to accelerate and diffuse to new parts of the globe. These breakthroughs occurred in very specific geographical locations that had significantly lower geographical constraints:

1. The emergence of Commercial societies in Europe about 800 years ago that innovated four of the Five Keys to Progress (productive agriculture, trade-based cities, decentralized power and export industries).

2. The diffusion of Commercial societies from Northern Italy to Flanders and then to the Netherlands and finally to Southeast England.

3. The migration of Europeans to much of the rest of the world. The migration of peoples from Britain to North America was particularly important.

4. The Industrial Revolution in Britain, which added the fifth Key to Progress (widespread use of fossil fuels). The Industrial Revolution involved the application of fossil fuels to transportation, communication, materials, agriculture and other technologies, increasing their usefulness. This dramatically increased the rate of innovation to a level where real progress could take place in Western Europe and North America.

5. The Allied victory in World War II, which ended the totalitarian threats of Nazi Germany, Imperial Japan and Fascist Italy. Once the economies of Western Europe and Japan recovered from the devastation of the war, they grew at an unprecedented rate for almost 30 years.

6. The collapse of the Soviet Union in the early 1990s, which ended the last of the great predatory empires and undermined the legitimacy of centralized political monopolies. As a result, country after country dismantled totalitarian and authoritarian regimes in favor of increased freedom, democracy and market-based competition.

> **KEY INSIGHTS**
>
> Six historical breakthroughs enabled progress to accelerate and diffuse to new parts of the globe.

These six historical breakthroughs enabled all Five Keys to Progress to spread through much of the world. The results are fully evident in the metrics that we examined in Chapter One.

I believe that the first of these developments, the creation of Commercial societies was the most important and also the least written about, so this chapter will focus on it. I hope to examine all six developments in greater detail in future books in this series.

The key development that made progress for the masses possible was the emergence of a new type of society sometime after 1200 in Northwestern Europe: Commercial societies. Commercial societies differed greatly from even the most complex Agrarian societies of Eurasia. These societies built on the relatively productive agricultural systems of Northwest Europe but channeled the food surplus towards autonomous trade-based cities full of free people with a wide variety of skills.

While Agrarian societies established government-sponsored monopolies designed to extract a food surplus to benefit political, economic and religious elites, Commercial societies promoted competition between a wide variety of institutions. The result was a dramatic increase in the rate of innovation of technologies, skills and organizations that triggered the first real progress in human history.

A brief note: I am cheating a little bit by considering Northern Italy as a part of Northwestern Europe. I do so because Northern Italy shares far more in common geographically with the Northwestern section of the European continent than the southern section, including a Temperate Forest biome, a large navigable river, stretches of productive Alfisol soil and summer rainfall.

Why Europe?

One of the most controversial and important questions in the history profession is why Europe developed so much faster than other regions on the planet. While historians disagree over exactly when Europe became the dominant economic and military region, none disagree that by the late 19th Century Europe dominated the globe. And even within the European continent, the much smaller sub-region of Northwestern Europe was the clear leader.

How did such a tiny portion of humanity play such a critical role in modern history?

By going down the list of geographical factors that we discussed in previous chapters, we can answer this question more systematically than other researchers have. Northwestern Europe has some distinctive geographical advantages over much, but not all, of the rest of the world.

To sum up, first of all, most of the European continent is covered by Temperate Forests, and the western edge (France, England and the Low Countries) is covered by Deciduous Forests. This fact alone gives Northwestern Europe a huge advantage over the vast majority of the planet. In addition, Northwestern Europe has large tracts of Inceptisol and Alfisol soils.

Northwestern Europe was also blessed with a large number of navigable rivers, including the Rhine, Seine, Meuse, Po and Thames. And virtually all of Northern Europe, except Norway and the Alps, has an altitude lower than 500 meters. Finally, the growing season of this region is unusually long. All of these geographical factors combined

> **KEY INSIGHTS**
>
> A few societies in Northwest Europe were able to overcome geographical constraints, escape the poverty trap and generate progress for their people. These societies evolved each of the Five Keys to Progress. When each of the Five Keys to Progress is in place, a society becomes a vast decentralized problem-solving network that promotes progress.

to make Northwestern Europe a perfect environment for complex Agrarian societies with productive agriculture.

The only geographical factor that could have made Northwestern Europe less than ideal was a lack of domesticable plants and animals. But the region's proximity to the Middle East erased this sole disadvantage over 5000 years ago. Whether they came by migration from the Middle East or trade, Europe acquired a wide variety of domesticated plants and animals. Without this proximity to the Middle East, Northwest Europe might have evolved societies similar to North America before the European conquest.

Because of these geographical advantages, it is difficult to see how complex Agrarian societies with productive agriculture would have failed to evolve in Northwestern Europe. But it is not clear that geography alone can explain why Northwestern Europe became much more influential than, for example, East Asia and particularly China. To explain this, we must go beyond geography.

I contend that what distinguished Western Europe from the rest of the world was a dense constellation of autonomous trade-based cities. These cities formed a society type like no other in the world: the Commercial society. While the earlier society types are difficult for a modern reader to imagine living in, Commercial societies are strikingly similar to our own. The only missing ingredient was the widespread usage of fossil fuels.

Commercial Societies

A Commercial society is a society whose members acquire the majority of their calories by using their skills to sell a product or service, so they can buy food from the marketplace. While the vast majority of inhabitants of the other society types acquired their food directly, many people living in Commercial societies depended upon others to do so.

Sometimes Commercial societies imported food from their immediate rural hinterland; sometimes they imported food from distant shores. While they did not have modern grocery stores as we do today, buying food in the marketplace makes them very similar to us.

Key technologies enabled Commercial societies to do the following:

- Transport people and cargo across vast oceans (using galleys and triple-masted sailing ships)
- Force elites to compete against each other peacefully (using inclusive institutions)
- Trade goods across long distances (using currency, banks and double-entry bookkeeping)
- Invest money in business enterprises (using joint-stock companies, bond markets and the stock exchange).

KEY INSIGHTS

Commercial societies survive by using their skills to sell a product or service, so they can buy food from the marketplace. Their key innovations were sailing ships, inclusive institutions, currency, banks, companies, bond markets and the stock exchange.

Commercial societies were crucial because they were the first society type that translated technological innovation into a higher standard of living for a large segment of society. In other societies, progress was only possible for the elites who acquired their wealth by extracting resources through the threat or actual use of violence.

Before Commercial societies, progress was zero-sum: one person could acquire a higher standard of living only at the expense of others. After Commercial societies, progress was positive-sum: one person gets progress by creating wealth. Creating wealth does not hurt other people in the way that extraction does.

The enrichment of one person in a positive-sum environment will most likely help others by creating a larger market for their skills. Commercial societies were fundamentally based on creating wealth through trade and innovation of technology, skills or social organizations, not expropriating wealth.

Preconditions of Commercial Societies

So why did Northwestern Europe evolve a network of autonomous trade-based cities, while other regions with Agrarian societies in Eurasia did not?

Geography played an important, but not a critical role. North-western Europe possessed a vast network of navigable rivers that flowed year-round and emptied into the ocean. The region also possessed a long, jagged coastline with natural ocean ports. But the same could be said for parts of China, Korea, Japan, India and some other locations in Eurasia.

While all previous transitions between society types was driven by geography and technological factors, whether or not an Agrarian society transitions to a Commercial society is mainly driven by political factors.

When an Agrarian society reaches a certain level of agricultural productivity, it opens up the possibility of transitioning to a Commercial society. Certain geographical factors, such as proximity to ocean ports or navigable rivers matter, but they are also common enough that most large societies have at least one geographic area that fits that criteria. This means that the key factor is political autonomy, or more accurately the ability of fledgling trade-based cities to defend and enlarge their political autonomy from aggressive Agrarian empires.

The lucky few cities that were able to carve out political autonomy evolved into full-fledged Commercial societies. The vast majority failed, becoming just another city within a larger Agrarian society, albeit with a more prosperous citizenry. Without political autonomy, the interests of trade were subordinated to the overriding importance of security and resource extraction within Agrarian elites.

In Northern Italy and the Low Countries, a constellation of autonomous Commercial societies was able to sustain themselves for centuries. Their demonstration effect on each other ("look, it is possible for a small city to go its own way") and temporary military alliances against neighboring empires allowed them to support each other.

Given the importance of Commercial societies in accelerating innovation of technology, skills and social organizations, the ability of fledgling trade-based cities to defend and enlarge their political autonomy from aggressive Agrarian empires is one of the great contingencies in world history.

Because a few regions on the European continent were able to make the transition, European culture and social organizations have had an out-sized impact on the rest of the world. Because no regions outside Europe were able to do so, the impact of the rest of the world after 1500 has been seriously circumscribed (at least until the last few decades).

> **KEY INSIGHTS**
>
> Commercial societies evolve when four of the five keys to progress are present (productive agriculture, trade-based cities, decentralization and export industries).

The Legacy of the Fall of the Western Roman Empire

As Walter Scheidel argued in his book *Escape from Rome*, Europe differed greatly from the rest of Eurasia in its formation of empires. While most of Eurasian history was dominated by a succession of empires, once the Roman Empire collapsed, European history was not. While the Franks, Carolingians, Spanish, French and Germans all tried to create continent-spanning empires, they all failed. Instead, Europe was fragmented into many competing kingdoms. This was a very different outcome from the rest of Eurasia.

After the collapse of the Western Roman Empire, Northwestern Europe experienced a radical decentralization of the political, economic and religious power called feudalism. While the rest of Eurasia evolved strong centralized governments with powerful extractive political and economic institutions, Northwestern Europe was highly fragmented by feudalism.

From the fall of the Roman Empire until the consolidation of nation/states in the 16th century, Europe was fragmented into tiny domains run by individual nobles or bishops. To give you an idea of how small these polities were, take a look on the map at the size of the few that still remain: Monaco, San Marino, Liechtenstein, Luxemburg and Andorra. Of course, there were larger Duchies and Counties, but few of them approached the size of modern-day states.

This political and institutional fragmentation enabled the rise of Commercial societies in Northwest Europe. Most cities and towns inhabited but a portion of these small political units. Sometimes by fighting for independence, but more often by bargaining with their lords peacefully, trade-based cities gradually carved out a level of political autonomy that was unheard of anywhere else in the world. Some of these cities evolved into independent polities, but most won a de facto autonomy on economic and political issues conditional on paying a communal tax to their lord.

> **KEY INSIGHTS**
>
> After the fall of the Roman Empire, no great empires dominated all of Europe. Instead, decentralized feudalism enabled Commercials cities to evolve.

Many lords found that by granting their cities and towns exemptions from feudal dues, they could generate greater revenues. Other lords offered autonomy to encourage settlers to move into their more sparsely settled regions. Because cities and towns created so much more wealth than even the most productive countryside, savvy nobles could make more money with a light touch.

Let me be clear. Commercial societies did not emerge in feudal realms. Commercial societies required free labor, while feudalism required serfdom.

But because feudalism was inherently decentralized, there were often small regions where serfdom and feudal lords were absent or weak. If those regions had geographical factors that favored the growth of trade-based towns, those towns had a real chance to grow into societies that were completely different from the surrounding feudal societies. Commercial societies evolved where both feudalism and centralized Agrarian institutions were weak or absent.

Geography of Commercial Societies

Just like all society types, Commercial societies could evolve only under highly constrained geographical preconditions. Most importantly, Commercial societies could only emerge where there was highly productive agriculture. This effectively meant that they could only emerge on the plains of Temperate Forest or Mediterranean biomes. Without a significant surplus of food, it would have been impossible to feed the citizens who lived in the cities that were essential to the survival of Commercial societies.

Commercial societies also needed to be located on a major navigable river or a natural ocean port. Before the advent of the railroad, the only cost-effective means of moving large, heavy goods was via sailing ship. This radically reduced the geographical areas that could potentially evolve into Commercial societies.

Commercial societies also had to be located in proximity to Agrarian societies. The elites of Agrarian societies who hungered to display their social status created a huge market for luxury goods. Since these luxury goods were small, light and valuable, they presented a relatively easy market to penetrate.

In Europe, this meant the spice trade with Southeast Asia. Once early Commercial societies in Italy learned how to make money by transporting and selling luxury goods, they could then branch out into less-profitable but more stable consumer markets, particularly textiles.

While Commercial societies needed to be close to Agrarian societies, they also had to be autonomous from their political and military power. Kings were notorious for looting wealthy cities with taxation or even direct expropriation. Kings also had a nasty habit of going to war to steal the wealth of weaker neighboring societies.

The decentralized political structure of feudal Europe was ideal for the growth of Commercial societies. Without one large empire to stifle these little cities, they could pop up like mushrooms wherever the local geographical and political conditions would allow.

Incubator of Commercial Societies

When one looks at a map of Europe, it is striking how many of the Commercial societies were located on the periphery of the Holy Roman Empire. The Holy Roman Empire lasted for many centuries and covered Belgium, Netherlands, Northern Italy, and most of modern-day Germany. Just a few of the trade-based cities that were within the borders of the Holy Roman Empire were Venice, Florence, Genoa, Bruges, Antwerp, Amsterdam, Cologne, Nuremberg, Augsburg, Lubeck and Hamburg.

This is not a coincidence. Among the empires of Europe, the Holy Roman Empire was extraordinarily decentralized. It did not even deserve the title of "empire".

The Emperor was elected, not inherited and many of the cities within its borders had very broad political and economic autonomy. Imperial cities, in particular, had so much autonomy that it was often a de facto independence. As long as they swore loyalty to the Emperor, they could do pretty much as they pleased. This meant that local merchant families, not kings or princes, were the de facto rulers of these cities and towns.

The outskirts of the Holy Roman Empire made perfect incubators for Commercial societies. The Holy Roman Empire ensured that local nobles, who swore loyalty to the Emperor, were not a military threat from within. The Holy Roman Empire also protected the cities from threats by rival Agrarian empires and barbarian attacks. But the Holy Roman Empire left them enough autonomy that they could focus on what they were good at: innovating and making money.

Modern-day Northern Italy, Belgium and the Netherlands all fulfilled these conditions and each evolved into complex Commercial societies. These Commercial societies concentrated artisans, merchants and other skilled workers into tight geographical proximity and gave them both the freedom and the incentive to use their talents.

Their need to import foodstuffs for survival created a strong incentive to their rural hinterlands to make local agriculture more efficient, which created a larger food surplus. Locals also had incentives to invest in improving the transportation network on both land and sea.

With relatively inclusive political and economic institutions, significantly less of this food surplus was extracted by elites, enabling incomes to grow. This created a virtuous cycle of innovation and growth that the world had never seen before to evolve.

Commercial societies could expand rapidly to other defensible coastal and riverine locations because they possessed complex nautical technologies. Because they had relatively small populations, however, they could rarely expand inland, particularly if that region was already inhabited by Agrarian societies.

> **KEY INSIGHTS**
>
> Commercial societies evolved in Northern Italy, then Flanders (present-day Belgium) and then the Netherlands.

Large volumes of trade and decentralized wealth enabled them to evolve inclusive political institutions. Rather than extractive institutions run for the benefit of a small political and economic elite, Commercial societies had decentralized institutions run for the benefit of a much larger segment of society (although rarely a majority).

Commercial societies, however, were terribly vulnerable to conquest from their far larger Agrarian neighbors. Though citizen militias and strong navies enabled them to hold out for centuries, the predatory Agrarian empires conquered many fledgling Commercial societies. Only those that were located on the fringes of the Holy Roman Empire tended to survive.

Pre-Industrial England

The most unusual Commercial society to emerge in the early modern era was England. While many historians believe that it was the Industrial Revolution that distinguished England from the rest of Europe, I believe that England and particularly Southeast England transformed into a Commercial society beforehand. Indeed, it was being a Commercial Society that enabled Britain to experience an Industrial Revolution in the first place.

By copying the innovations from other Commercial societies, particularly Flanders and the Dutch Republic, Southeastern England effectively became a Commercial society within a larger Agrarian England. Southeastern England possessed the innovativeness of Commercial societies, but with a far larger population base.

In some ways, England was little different from the Agrarian societies in both Europe and Asia. England had a centralized monarchy, titled elites with massive landholdings and limited urbanization. This made England very similar to Agrarian societies.

KEY INSIGHTS

Southeast England copied technologies, skills and organizations from Flanders and Netherlands to become its own Commercial society.

However, unlike most of the rest of the European continent, major portions of England gradually transitioned from being an Agrarian society to being a Commercial society. This transition enabled a level of economic growth that differentiated England from most of the Continent long before the Industrial Revolution.

This transition occurred for a few key reasons. First, Britain had a huge geographical advantage: the English Channel. The "world's largest moat" was a substantial geographical barrier to Medieval armies who lacked organized naval power. Of course, this made it difficult for England to project power onto the Continent as well. Unlike the Commercial societies of the Continent who were constantly fighting for survival against larger Agrarian states, England had far greater security.

While England shared the geographical advantages of Northwest Europe — Temperate Forest biome, productive soils, extended plains and navigable rivers — the other societies on the British Isles were not as blessed. Wales, Scotland and Ireland were largely highlands with less productive soil, and they were much further from the dynamic economies of Flanders and the Netherlands. Because of this, the other societies in the British Isles were far poorer and weaker. For these reasons, England rarely faced a serious military challenge within the British Isles.

This lack of a clear military threat meant that England never needed the huge standing armies that the Continental powers found necessary to survive. While kings on the Continent had to fight to survive, the English monarchs could largely choose their wars, and if they lost, the entire kingdom was generally not in danger.

This meant that the aristocracy and the monarchy gradually worked out a balance of power that maintained a centralized government that protected society from hostile external powers while respecting the economic rights of citizens. This balance of power was institutionalized in the House of Commons and English Common Law. While not as inclusive as governments in other Commercial societies, England's political and economic institutions were far more inclusive than the Agrarian states on the Continent.

Proximity to the dynamic Commercial societies of Flanders and Netherlands was central to the transformation of England from an Agrarian society to a Commercial society. There was a continual flow of people, technologies, skills, social organizations and values from the Low Countries to Southeast England. While the transformation was much slower to diffuse to other parts of England, East Anglia and London played a leading role.

Innovation and Progress in Commercial Societies

Commercial societies had many similarities to the modern societies of today. They were based upon large, autonomous trade-based cities with large numbers of artisans, traders, merchants and bankers. The majority of the citizens acquired their food by purchasing it in the market. Many of those citizens were highly skilled in emergent technologies of their day (what we would now call "cutting-edge" technology). Commercial societies had very high levels of urbanization and their citizens were highly literate.

Because of those characteristics, Commercial societies were extraordinarily innovative. They made important innovations in agriculture, energy, transportation, industry, finance, education, military, urban services, politics, science, art, diplomacy and many other fields. Indeed, it is hard to think of a domain in which Commercial societies did not make important contributions.

With their dense concentrations of free citizens with widely varying skills, Commercial societies were technology aggregators. Commercial societies copied the best technologies, skills and social organizations from other commercial societies. Then they recombined those innovations in new ways to create even more technologies, skills and social organizations. Each city tended to focus its resources on a few potentially profitable industries and innovate in those areas. This built the foundation for exporting technologies. Then other cities could copy those innovations and modify them for different purposes.

Copying other societies came naturally to Commercial societies. Their day-to-day workings naturally promoted frequent contact between trading ships, traveling merchants, skilled craftsmen and ship crews. Many Commercial societies had large numbers of skilled immigrants who brought new ideas and skills to their adopted homeland.

> **KEY INSIGHTS**
>
> Commercial societies are very innovative and willing to copy the innovations of other societies. Nothing like them had ever existed.

Commercial societies naturally increased the productivity of agriculture (the first Key to Progress). Because the citizens of Commercial societies purchased their food in the market, this gave a strong incentive for local farmers to increase their food production and specialize.

At first this impact was only in a small hinterland within walking distance of the city, but it gradually spread outward. The more the population of the city grew, the more local farmers were drawn into producing for the market. As more and more farmers were drawn into the market, the productivity of agriculture gradually increased until it far surpassed neighboring Agrarian societies.

Because commercial transactions required a certain amount of trust in strangers, Commercial societies tended to have much higher levels of trust than Agrarian societies. This helped them adopt a positive-sum mentality where strangers can cooperate to the benefit of both parties. This was a distinct change from the zero-sum mentality that dominated Agrarian societies.

Commercial societies had a strong incentive to innovate and copy new transportation and communication technologies. Being linked to each other via galleys and later sailing ships, each city could copy what worked best from all other cities, leading to far higher levels of innovation than in the surrounding Agrarian societies. As more and more Commercial societies evolved and their total population grew, innovation slowly accelerated.

Each Commercial society acquired technologies, skills and innovations from the other Commercial societies. This ensured that the process of innovation was additive. Breakthroughs were never lost, and the useful ones were always built upon.

With each generation the population of Commercial cities grew larger and larger, accelerating the recombination of old ideas to form new ideas. This ensured that innovation and progress were constantly accelerating.

Perhaps the most important innovation of Commercial societies was inclusive institutions. While centralized extractive institutions dominated Agrarian societies, elected mayors and town councils ran Commercial societies. Local militia filled with free citizens defended the cities from invaders. Guilds competed with each other for economic gain and political power. Many cities had a variety of religious institutions competing with each other for worshipers.

Most importantly, each of the Commercial cities competed against each other. Even if a small group of merchants or guilds could control one city, they could not control all Commercial cities. Any centralized institution that undermined innovation would cause the city to decline and another city to rise up to take its place.

It was this decentralization of political, military and religious power (the third Key to Progress) that enabled progress to get started in Commercial societies. Success was no longer about climbing up to the top of centralized monopolies. It was now necessary to out-innovate the competition.

As Commercial societies grew in number as well as population, their people grew richer. As they grew richer, its people could afford to pay others to solve their problems. At first the problems were about basic survival, but gradually they shifted to paying for a higher quality of life. As merchants, artisans and other skilled workers competed with each other to offer solutions in the marketplace, progress began to take hold.

Commercial societies found ways to increase cooperation to solve common problems. Cooperation in previous societies was largely based upon reproduction, self-defense and food production. Commercial societies added cooperation within the marketplace to solve virtually any problem that others were willing to pay money for.

By the 16th Century, Commercial societies were surging away from the rest of the world in the complexity of their technologies, skills and social organizations. More importantly, they were starting to produce real progress that changed a person's standard of living within one lifetime. Progress had evolved.

Indeed, between 1200 and the Industrial Revolution these societies accounted for the bulk of the innovation and all of the progress, making Europe by far the most innovative continent during this period. Without these Commercial societies, however, Europe would have been little different from Asian Agrarian societies in their rate of innovation.

Economic Growth in Commercial Societies

All of these innovations led to the strongest economic growth the world had seen up to that time. Estimating the wealth of societies that existed centuries ago is not easy. Fortunately, Angus Maddison has created his extremely useful database with estimates of per capita GDP for past societies. His figures are, of course, just estimates. At least for now, however, they are by far the best method for determining the wealth of past societies.

When we use Maddison's estimates of historical per capita GDP, it is clear that Commercial societies experienced a different growth trajectory from other societies. In the year 1000, all complex societies in Eurasia had a per capita GDP between $400 and $650/year (i.e. at or just above subsistence levels).

It seems likely that this is roughly the same standard of living that humans experienced for all of history before that year. There is no evidence that any society reached a higher level before 1000, except in a few Imperial capitals that acquired their wealth by extracting it from other regions, which would have had little to no impact on the median standard of living. Progress was something humans of that era and previous eras could simply not comprehend.

By 1500 things were beginning to change. While all the Asian Agrarian societies were stuck at subsistence levels, the European Commercial societies experienced an upturn in per capita GDP. This is likely the first sustained increase in economic growth in world history.

More importantly, Italy jumped up to $1,100/year. This was a 144% improvement over its levels in 1000. And since Northern Italy was certainly more prosperous than Southern Italy, the actual level of economic development of the Northern Italian city/states was certainly much higher than $1100/year. One can also see a noticeable upturn in Flanders (in present-day Belgium) from $425 to $875/year.

After 1500, we see an upturn in both the Netherlands and the United Kingdom. The United Kingdom went from $714/year in 1500 to $974/year in 1600. This made the United Kingdom the third richest society in the world at the time. And since Southeast England had a higher standard of living than Scotland, Ireland and the rest of England, levels in Southeast England were certainly even higher.

After 1500, both the Italian and the Flemish economies failed to keep up. Italy experienced almost no economic growth in the next 320 years, while the Flemish economy grew at the rate of the rest of Europe. The military occupations of the Spanish, French and Austrian empires took a huge toll on both societies.

The best evidence of progress, however, was in the Netherlands, which grew from $761/year in 1500 to $1,381/year in 1600. This was probably the fastest sustained economic growth up to that time, and it firmly established the Netherlands as the richest society in the world.

Dutch economic growth continued through most of the 17th Century, reaching a level of $2,130/year in 1700. That is the highest per capita GDP that any nation would reach before the Industrial Revolution.

And given that detailed economic analyses conclude that the Dutch economy peaked around 1670, this figure may understate the actual peak level. It is not a coincidence that 1670 marks the year when Agrarian France invaded the Dutch Republic and almost destroyed the regime. Once again we see evidence of invasion by Agrarian empires

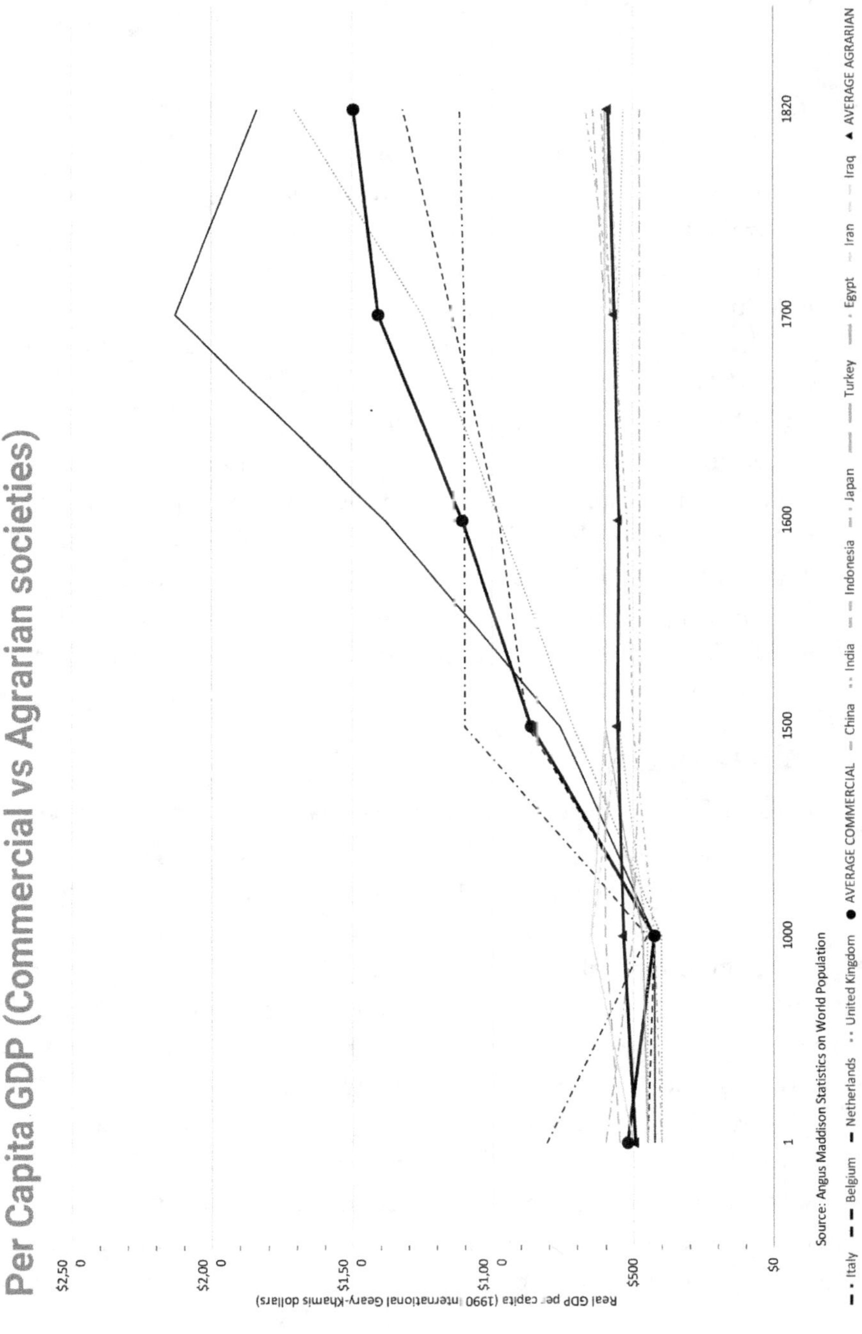

Per Capita GDP (Commercial vs Agrarian societies)

Source: Angus Maddison Statistics on World Population

- - Italy — - Belgium — Netherlands ·· United Kingdom ● AVERAGE COMMERCIAL ·· China ·· India ·· Indonesia ·· Japan — Turkey — Egypt — Iran — Iraq ▲ AVERAGE AGRARIAN stifling much smaller Commercial societies.

After 1700 the United Kingdom and the Netherlands moved in opposite directions. The Netherlands declined in the 18th Century until it fell to $1,838/year in 1820. Some of this was undoubtedly due to the French occupation during the Revolutionary and Napoleonic wars and their trade blockades, but it is clear that the Dutch economy was in decline well before that time.

Meanwhile, the economy of the United Kingdom grew from $1.250/year in 1700 to $1,706/year in 1820, approaching the level of the Netherlands. It is important to note, however, that the British rate of growth during the 18th Century was almost identical to the rate of the 16th Century, overturning the commonly held belief that it was solely the Industrial Revolution of the 18th and 19th Century that transformed Britain's fortunes. Though cotton textile factories transformed parts of Northern England and Lowlands Scotland, their overall effect on the standard of living in the rest of Britain before 1830 was very small.

European Settlers

As James Belich argued in his book *Replenishing the Earth*, an important historical breakthrough that helped to spread progress to new regions was the migration of Europeans to much of the rest of the world. European settlers brought their genes, technologies, skills, social organizations and values with them. The result was often a dramatic jump in per capita GDP in local regions. This jump rarely benefitted the indigenous people, at least in the short run, but it did lay the foundation for widespread progress outside Europe.

Particularly important was the mass immigration of the British to North America. Remember from earlier chapters that North America had almost all the geographical preconditions for building complex Agrarian societies. The main constraint was a lack of domesticated animals capable of pulling plows.

Because many of the early British immigrants came from complex societies, the settlers brought domesticated animals with them. They also brought an entire suite of complex technologies, skills and social organizations. Just as importantly, they did not bring the centralized extractive institutions from their home countries.

KEY INSIGHTS

European settlers, particularly British settlers to North America, brought the technologies, skills and organizations that made progress possible outside Europe.

When settlers arrived in a relatively unpopulated region, they tended to recreate the societies that they had left and to adapt this to the local environment. It could not be otherwise.

Settlers had grown up in societies with a unique suite of technologies, skills, social organizations and values. They could not reinvent themselves as people just because they were living in a new land. Their survival and reproduction were at stake. The difficulties of the initial settlement forced them to fall back on what they already knew.

Settlers from the largest Agrarian societies of Western Europe — Spain, France and Portugal — tended to recreate similar societies in the New World. They all left hierarchical agricultural societies that highly valued status, military glory, conspicuous consumption and avoidance of physical work. It was only natural that the conquerors would want to set up similar societies in the New World, but with themselves at the top of the pyramid.

If similar people had settled all of North America, it too would have been sculpted into similar societies with all the disadvantages of Agrarian societies. Fortunately for humanity, a very different type of people settled in what became the Northeastern United States. Settlers in this area largely came from Commercial and Free Peasant societies.

The establishment of rapidly growing Commercial and Free Peasant societies in North America by European settlers sowed the seeds for a fundamental shift in global power. As the United States grew in population and geographical extent, industrialized and finally built a world-class military, the balance of world power between Agrarian societies and Commercial societies fundamentally shifted.

No longer were free societies a small collection of city-states huddling on the fringes of Europe. Now they were continental in scope. This development played a fundamental role in shaping the history of the 20th Century.

KEY INSIGHTS

In the 19th Century, Britain created the first Industrial society. Industrial societies are much like Commercial societies, except they use vast amounts of fossil fuels.

Industrial Revolution

We have seen how Commercial societies in Northwest Europe created the first self-sustaining economic growth that led to real progress for the masses. They did so because they possessed four of the Five Keys to Progress (productive agriculture, trade-based cities, a decentralization of power and exporting industries).

But the lack of the fifth key — widespread use of fossil fuels — placed a fundamental cap on how far that progress could continue. The Dutch Republic in 1670 with a per capita GDP of $2,130/year appears to have hit this cap.

It was not for another 176 years that another nation was able to break through this limit (the UK in 1846). What followed was another 176 years of economic growth and still counting. Most importantly, this economic growth spread beyond a relatively small network of Commercial societies to encompass much of Europe and North America.

The United Kingdom achieved this result by pioneering a new society type: the Industrial society. The Industrial society is very similar to Commercial societies in that people acquire their food by selling a skill or product in the marketplace so that they can purchase food within that same market. The key difference is that the United Kingdom combined this with the widespread use of fossil fuels.

Key technologies enabled Industrial societies to do the following:

- Exploit the awesome energy density of fossil fuels
- Dramatically increase the productivity of agriculture (using tractors, nitrogen fertilizer, etc.)

- Build technology that is so cheap that even the poor can afford to purchase it (using assembly lines)
- Transport people and goods rapidly, cheaply and across long distances (using railroad, steamships, automobiles, airplanes and container ships)
- Communicate across long distances (using the telegraph, telephone, radio, television, mobile devices and the internet).
- Develop new materials and mass produce them for sale (steel, concrete, asphalt, glass and plastic)

In pre-Industrial societies, the vast majority of useful energy was derived from human muscle power, animal power, burning wood, or harnessing water or wind. All of these power sources ultimately derived from the sun. Industrial societies added on energy derived from fossil fuels, primarily coal. In the later phases of the Industrial Revolution, petroleum and natural gas played important roles.

The use of fossil fuels injected a huge amount of useful energy into Industrial societies. They found new ways to apply this energy to transportation, communication, materials and agriculture. The railroad, steamship, factory, electrical grid, telegraph, radio, steel, nitrogen fertilizer, automobiles, trucks, container ship and airplane are just a few of the innovations that helped to sustain progress.

Most importantly, these innovations dramatically increased the material standard of living for the masses.

As an aside, it is important to note that the use of coal for heating homes was common in London and some other English cities long before the Industrial Revolution. No other society used coal to anywhere near the extent of pre-Industrial England.

I do not believe that this undermines my overall argument as heating homes is a specific application of fossil fuels with limited benefits to other sectors of the economy. Warm homes are far more comfortable to live in, but they do not transform societies as the application of fossil fuels to agriculture, transportation, communication and manufacturing do.

I do believe, however, that pre-Industrial coal usage for heating homes gives us an important clue as to why the Industrial Revolution took place in Britain. The use of coal gave Britain critical technologies, skills and organizations that could later be modified and scaled up. While they did not know it at the time, this gave Britain a critical skills advantage over other Commercial nations that might have industrialized earlier.

Some Societies Copy the UK

The Industrial Revolution in the United Kingdom created benefits that were obvious to much of the rest of the world. In particular, the economic benefits enabled the British Empire to reach new heights of political and military power. Some nations resented this power, while others chose to copy the technology, skills and organizations that made it possible. Those that were successful in copying experienced rapid and sustained economic growth.

Nations that had previously been Commercial societies found it much easier to copy the UK as they had already built four of the Five Keys to Progress. Belgium's per capita GDP increased by 67% between 1846 and 1872. The United States' per capita GDP increased by 282% from 1850 to 1929. Canada's per capita GDP increased by 172% from 1878 to 1913. Australia and New Zealand experienced even faster growth rates.

Nations that had previously been Free Peasant societies were also able to adopt Industrial technologies, skills and organizations fairly easily. Though they lacked the cities of Commercial societies, their egalitarian social structure and far more inclusive political institutions gave them a strong foundation to build upon. Switzerland, Denmark, Sweden and Norway increased their per capita GDP by between 88% and 270% in relatively brief periods.

Agrarian societies were far more resistant to adopting Industrial technologies, skills and social organizations. The political, economic and religious elites who controlled extractive institutions saw Industrialism as a potential threat to their power. There was one domain, however, where the elites could not ignore the effects of industrialization: military power. Industrial Britain had fundamentally disrupted the global balance of power, and the Agrarian powers could not accept the idea of falling too far behind.

Was The Industrial Revolution Inevitable?

This leads to an interesting question. Is modern progress an inevitable outcome given the human ability to innovate new technology? Is it just a matter of time? Do all the factors that I discuss in this book obscure a much simpler process?

These questions are important because they change the way that we view the origins of modern progress. If this were true, then it would seriously weaken my argument that the Five Keys to Progress and Commercial societies are critical concepts through which to understand modern progress.

In an influential article called "Was an Industrial Revolution Inevitable? Economic Growth Over the Very Long Run," Charles Jones develops a simple mathematical model that he believes explains modern progress. It takes into account all the observations that I made earlier about innovations being the recombination of existing ideas or technologies. He shows that the nice curve predicted by his model closely resembles long-term economic growth.

Jones' model predicts long periods of very slow economic growth followed by a sudden surge. This sudden surge appears to "come out of nowhere," but his model shows that it was just a continuation of the same process over long periods. It was the inevitable outcome of exponential growth caused by recombination. His interpretation is that the Industrial Revolution was inevitable given enough time.

I love the model, but I disagree with the interpretation because it neglects other factors. Technological innovation is not the only factor in human progress. The rates of innovation are very sensitive to conditions. Innovation requires food. Food is not inevitable. Babies consuming the food surplus played a major role in slowing technological innovation.

Nor does an increased population inevitably increase the rate of innovation. In traditional societies, very few people live in cities, typically around 3% of the population. Farmers create very few technological innovations outside the agricultural sector.

It is cities where the action takes place. For this reason, the increase in the population of cities is far more important than the overall increase in population. And even among city dwellers, it is at most a third of the urban population who systematically develop the skills needed to innovate.

Innovation also requires that elites do not stifle the transfer of food surplus to cities where people of skills live. This makes the relative autonomy of trade-based cities essential to innovation.

In Agrarian regimes, elite expropriation drastically reduced the rate of technological innovation and the material benefits that went to the masses. Only Commercial cities were able to combine plow-based agriculture and reduce elite expropriation. Both were essential to technological innovation.

Nor is there a simple relationship between the size of the urban population and the number of skills. Small cities have a relatively low diversity of skills even if they are autonomous and trade-based. The bigger the city, the greater diversity of skills. Based upon what we know about networks, it seems likely that this relationship is also exponential.

Plus it matters which skills a city has. In any one time period after the evolution of Commercial societies, there was a very small number of cities on the leading edge of technological innovation. Those cities were

> **KEY INSIGHTS**
>
> Industrial technologies, skills and social organizations enable people to overcome many, but not all, of the geographical constraints that previously undermined the possibility of progress. For the first time, people anywhere on the globe had the opportunity to experience progress. They could now escape the trap of geographical constraints.
>
> Soon after the Industrial Revolution, Western Europeans and Americans copied the industrial technologies, skills and organizations from Britain. This rapidly spread progress to their people.

Venice, Bruges, Antwerp, Amsterdam and London in temporal order.

These cities combined the technologies, skills and organizations that were most relevant to the newest innovations and the ones that had the biggest impact on societies. Due to what economists call "agglomeration" effects, entrepreneurs, artisans, skilled workers, organizations and capital all concentrated in this one city. New York City, Chicago, Detroit and Silicon Valley played similar roles in later eras.

Charles Jones' model also does not account for the fact that human societies can be very different from each other in the amount of technology and the rate of innovation despite its members being from the same species. As we already saw, geographical constraints placed much harder limits on some societies than others. As those societies gradually innovated technologies and approached their geographical limits, those societies diverged radically from each other in the level of wealth.

It was only when a society acquired the Five Keys to Progress that crossing the threshold to modern progress became possible. Geography, demographics, limited food, energy and politics all placed very real constraints on both the rate of innovation and the degree to which innovation could lead to benefits for the masses.

Without Commercial societies, which invented the first four Keys to Progress and the existence of fossil fuels, the Industrial Revolution would not have been possible. And given how many Commercial city/states were conquered by Agrarian regimes that stifled their progress, it is entirely possible that all Commercial societies could have been wiped off planet Earth.

During the Commercial era, these cities were highly vulnerable to military invasion. Antwerp's economy was forever disrupted by the brutal occupation by the Spanish in 1576. Amsterdam's economy suffered similar, though less fatal, disruption from the French invasion of 1672 and later wars with Britain. London's economy might also have been seriously interrupted if the Spanish Invasion of 1588 had ended differently. The Normans in Southern Italy as well as the Spanish, Austrian and French empires conquered many other smaller, but promising Commercial city/states.

Because the Commercial societies emerged in very specific geographical and political conditions that were unique to Northwest Europe, it is safe to assume that their destruction would have forever eliminated the possibility of the Industrial Revolution.

So, no, I do not believe that the Industrial Revolution was inevitable. The Industrial Revolution required all Five Keys to Progress to come together at one time and place. The place was Northwest Europe and the time was between 1200 and 1830. The story could have worked out very differently.

> **KEY INSIGHTS**
>
> The Industrial Revolution was only possible because of Commercial societies, and they were not inevitable. Without Commercial societies, modern progress would not be possible.

Great Power Conflicts of the 20th Century

Another potential turning point in history was the great political and military confrontations of the 20th Century: World War I, World War 2 and the Cold War. The combined military power of the United States and the British Empire was essential for outcomes that did not disrupt the spread of progress.

World progress could probably have survived and continued with a German victory in World War I, but an Axis victory in World War II or a Soviet victory in the Cold War would likely have been catastrophic. While the Nazis and Communists liked the technologies that free societies produced, their political and ideological model was fundamentally hostile to the decentralization of power and open experimentation. One shudders to think of the consequences of failure in these global struggles.

British and American military power in the 20th Century came directly from the fact that they were early Commercial societies and early industrializers. The technology, skills and organizations that came from Commercial societies and the Industrial Revolution did not just enhance people's material standard of living. They also led to a wave of innovations in weapons and armor. Perhaps more importantly, industrial innovations to energy, agriculture, transportation and communications technology all had important military applications.

Soldiers need food. That food must be grown, harvested, processed, preserved and distributed so that soldiers can consume it on the battlefield. Indeed, food had always been a critical constraint on military power and strategy. One of Napoleon's most famous remarks, "An army marches on its stomach" is as true today as it was in his times. Industrial agricultural and transportation technologies made huge industrial armies possible.

Modern tanks, fighters, bombers, battleships, aircraft carriers, submarines are all weapons platforms derived from industrial transportation technologies. The massive consumption of ammunition and fuel requires vast supply chains that make industrial transportation technologies necessary. The telegraph and radio are all industrial communication technologies that enhance military command-and-control. All of these technologies require vast consumption of fossil fuels, manufacturing and industrial materials.

All of this means that the Industrial Revolution radically shifted the global balance of power. This caused the spread of progress to the rest of Europe and Japan.

Agrarian leaders, who could not care less about the material standard of living of the peasants, did really care about the military balance of power. Military power and violent competition were central to their perception of the world.

What had previously been conservative monarchical regimes concluded that they needed industrial technologies to maintain their relative military power and gain a competitive advantage on late industrializers. They knew that this would inevitably cause great, potentially destabilizing changes, to the very fabric of their societies. They even knew that those changes might threaten their power. In previous eras, they would not have considered "reform from above", but now they knew that it was essential for survival.

Three Agrarian powers — Germany, Russia and Japan — attempted to copy the technologies, skills and organizations of Western nations without copying their decentralized and competitive political institutions. They effectively tried to meld together the extractive institutions of Agrarian societies with an Industrial society. In doing so, they turned their nations into industrial and military powers of a very illiberal kind. These nations played a fundamental role in the wars of the 20th Century.

Because many of the rulers of these societies were born before industrialization hit their nations, they preserved the fundamentally zero-sum thinking of their Agrarian heritage. They saw military conquest and expropriation as the means to national progress rather than consensual cooperation and trade.

In doing so, the "reformers from above" sowed the seeds for revolution and radical ideologies. Agrarian regimes in Europe simply could not survive the brutal competition of World War I. Whereas the West, with its long Commercial heritage, was far more urbanized and had far more productive agricultural sectors, the other powers did not have those advantages.

Nations that had never gone through the transition to Commercial societies had smaller urban populations and less productive agricultural sectors. This limited the number of soldiers because they still needed large numbers of farmers to create the necessary food surplus. It also limited their ability to manufacture weapons and ammunition. In the long drawn-out confrontation of World War I, they finally cracked. The Russian empire collapsed in 1917, while the German, Austro-Hungarian and the Ottoman empires collapsed in 1918.

Worse, it was not just military defeat. It was a collapse of the entire Agrarian societies. While their leaders believed that they could meld industrial militaries onto Agrarian societies, this was not possible under the pressures of total war. Most likely, if World War I had never taken place, the European powers and Japan would have made a long, slow transition similar to current developing nations. But total war demands rapid change and failure to adapt drives entire regimes into extinction.

Rise of Totalitarian Ideologies

Taking advantage of the sudden collapse of the Tsarist regime in 1917, the Communists invented a new type of society. Then the Fascists, Nazis and Japanese militarists copied the Communists in some respects. These ideologues attempted to meld together Industrial economies with totalitarian governments that were far more extractive than even the worst Agrarian regime. They did so not to enrich elites (although the leaders of the Communist party lived at a far higher standard of living than the masses), but to fund the construction of a utopian society.

For much of the 20th Century, it looked as if these types of societies were destined to dominate the world, and the Industrial societies with a Commercial heritage (namely the United States and Britain) would fail. But Nazi Germany, Fascist Italy and Imperial Japan destroyed themselves with an orgy of reckless military expansion. The radical centralized power of those regimes could not restrain itself from creating more and more enemies. Decentralization of power might have restrained these self-destructive impulses, but this was not possible within this type of regime.

The Allied victory over Nazi Germany, Fascist Italy and Imperial Japan at the end of World War II appeared to be a fundamental setback for these types of societies. Unfortunately, the Soviet Union and its Communist alternative society benefitted as much as the western Allies.

For many decades, it looked like Communism was the wave of the future. The rigid control of totalitarian governments appeared to have huge advantages over the decentralized power of Western capitalist societies.

Even some non-Communist leaders saw the Soviet Union as a model to build upon. But the fundamental inability of centralized institutions to innovate, adapt and copy caused these regimes to implode in the long run. The Soviet economy destroyed itself slowly due to its economic contradictions.

The fundamental problem with the totalitarian regimes of the 20th Century was that they tried to avoid implementing the third Key to

Progress (decentralization of power). They had no problem with the other four keys, but because of their ideologies, they could not see the advantage of non-violent competition between political, economic and ideological organizations. They believed that their ideology unlocked the door to a utopian society. With such a goal, competition and diversity of views would only get in the way.

The Nazis and Fascists did not realize that decentralized political power could restrain a society from constantly going to war. With each new war, they created more enemies. Eventually, their enemies grew numerous and became so powerful, that defeat was inevitable.

> **KEY INSIGHTS**
>
> Modern totalitarian regimes seek to use industrial technologies without the third key (decentralization of power). This dooms them to failure in the long term, but it makes them dangerous military adversaries in the short term.

The Soviets and later the Chinese Communists did not realize that privately-owned companies competing against each other works far better than centralized planning. And centralized economic planning makes competitive export industries (the fourth Key to Progress) very unlikely.

Totalitarians always believe that decentralized power and non-violent competition between political parties, corporations and ideologies are weaknesses of a decadent society. They will never understand that decentralized power and non-violent competition are actually strengths.

And the failure to understand this inevitably dooms them to long-term failure. Unfortunately, the failure to learn this lesson early enough led to the tragic death of tens of millions of innocent people.

Progress Since 1991

The most recent breakthrough in the spread of progress for the masses was the fall of the Soviet Union in 1991. This event delegitimized the radical centralization of power and utopian ideologies that are central to totalitarian regimes. And it also removed forces that had kept small Communist elites in power in many nations.

Without Soviet occupation, military and economic aid and expertise in constructing brutal security apparatuses, leftist regimes collapsed one after another. With the obvious exceptions of China, Vietnam, North Korea and Cuba, it was an almost total wipe out of Communist regimes.

KEY INSIGHTS

Since the fall of the Soviet Union in 1991, we have seen the greatest spread of progress that humanity has ever known.

Though Communist China has made an epic comeback to world power, the regime is every bit as capitalist as the rest of the world. Communist China has effectively melded together a capitalist economy with a totalitarian political order that allows just enough freedom to encourage economic growth, but not enough to threaten the political power of the Communist party.

While the transition has been politically controversial, nation after nation has slowly learned to copy the technologies, skills and organizations of the Western Industrial societies. The result has been the dramatic progress documented in Chapter One.

Without Britain and the United States showing the way this could be done, none of this would have been possible. The earlier industrializers went through the long process of learning how to innovate technologies, master the new skills that those technologies required and founding new social organizations to maximize their usefulness. They also radically lowered the cost of manufacturing those technologies at scale, making them affordable to the poor.

More closely related to the metrics that we discussed in Chapter One, the people learned to adapt those technologies, skills and social organizations to improve the quality of sanitation, literacy, drinking water, health care, immunization, housing, food, education, and lowering the levels of violence, war and forced labor.

What had once been a small network of trade-based cities in one corner of Eurasia has spread to billions of people across the planet. Some of the nations that industrialized earlier innovated new solutions to their problems. As they worked out solutions, largely by trial-and-error, they created models that could be copied by countries that were previously locked in poverty by their geography.

The Great Problem-solving Network Expands

What were are seeing with each of these six historical breakthroughs is the gradual spreading of the Five Keys to Progress across the world. When each of the Five Keys to Progress is in place, a society becomes a vast decentralized problem-solving network.

Modern societies are composed of millions of technologies, skills and social organizations that humans link together in creative ways to solve problems. Many of these problems are so mundane that we take them for granted. Eating food, drinking clean water, disposing of waste, getting to work, seeing at night, and communicating with friends, family and coworkers.

Citizens in those networks freely join together in social organizations to combine technologies and skills to solve other people's problems. They typically do so not out of a vague sense of wanting to help other people, but to earn enough money to survive, reproduce and enjoy their brief time on earth.

Progress comes from decentralized networks of technologies, skills and social organizations. Organizations within these networks compete with each other to survive. That competition forces the members of those organizations to cooperate more closely, adopt the latest technologies, learn the latest skills and adopt the latest processes. All of this drives progress.

A key factor in the success of organizations and individuals is the ability to copy what works. We often tend to think of innovation as creating something entirely new, something that no one else could even imagine.

Innovation in the real world is often far more mundane: it comes from copying technologies, skills and processes that have already been proven to work, combining them with other existing technologies, skills and processes and then modifying them to function optimally in the new environment.

The real key to innovation is not coming up with a new idea or building the first prototype. It is the long process of diffusion and implementation that occurs as the innovation gradually spreads through society. Ironically, as we have seen, copying is the key to innovation.

Conclusion

Because of their unique geographical and political environments, a few societies in Northwest Europe were able to evolve into Commercial societies. For the first time, this enabled societies to evolve four of the Five Keys to Progress. This radically accelerated their rate of innovation of technologies, skills and social organizations. It was now possible for people to enrich themselves by creating wealth instead of extracting it from others.

European settlers carried Commercial technologies, skills and social organizations around the world. In particular, the settling of North America enabled a rapid population increase for societies based upon commercial principles. No longer constrained to a small network of trade-based cities, the United States was almost continental in scope.

The Industrial Revolution in Great Britain added the fifth Key to Progress: widespread use of fossil fuels. The enormous energy density of fossil fuels radically expanded the rate of innovation. Key breakthroughs in food, materials, transportation and communication technologies vastly increased the standard of living for the masses in Great Britain, Northwest Europe and North America.

The Allied victory in World War II and the Western victory in the Cold War broke the threat of totalitarian powers to the emerging progress. Particularly following the collapse of the Soviet Union in 1991, societies throughout the world began to copy the innovations of Industrial nations. The result has been the most radical expansion of the standard of living of the masses in all of human history.

WHY WE IGNORE
PROGRESS

So far we have seen that most traditional societies could not experience progress because of fundamental geographic constraints on food production. This created a trap that made progress impossible for millennia. A few societies in Northwest Europe evolved the Five Keys to Progress, thereby enabling the masses to escape the poverty trap and experience progress for the first time in human history. Gradually, over the past few centuries, that progress spread until it has encompassed the vast majority of mankind.

Based on the evidence that I have presented in previous chapters, one might expect everyone to be well aware of the progress that surrounds them and to have a positive outlook on the future. After all, every metric that I presented to document progress in Chapter One is easily accessible on the internet.

Unfortunately, this is far from the case.

Ignorance of Progress

Hans Rosling wrote a gem of a book, entitled *Factfulness: Ten Reasons We're Wrong About the World — And Why Things Are Better Than You Think*. I strongly encourage you to read it.

Rosling created a questionnaire consisting of 13 fact-based questions about the world. Rosling has handed out this questionnaire to college-educated people throughout the world on repeated occasions. There were no trick questions. Anyone with a basic knowledge of how the world is could have got most of the answers correct. But almost no one got the answers correct a majority of the time.

Typically, people choose the correct answer about 10-20% of the time. Almost no one got eight or more questions correct. The vast majority of people (80%) gave less correct answers than they would have achieved by randomly guessing!

And the better educated a person was, the fewer correct answers they got. The only question where the majority of the people got the answer correct is about climate change, a topic outside the scope of this book.

What is going on here? Perhaps Hans Rosling had too small a sample size, or he wrote biased questions. After all, he is not a professional pollster. Presumably, a professional polling organization would get better results, right?

Wrong!

In 2017, Ipsos teamed up with the Gates Foundation to take a survey across 28 different nations about people's views on progress and expectations for the future. The primary purpose of the poll was related to development aid, but many of the questions are relevant to perceptions of progress.

Ipsos concludes: "Among the findings from our 28-country study are:

- Most citizens in donor countries believe that living conditions in the developing world are worsening when most data... shows marked progress...
- Furthermore, few people in donor countries expect the quality of economic opportunities, health, or education in the world's poorest countries to improve over the next 15 years; ..."

However, among respondents who are best informed about

development progress, optimism increases.

What I found most striking about the results was how negative respondents from wealthy countries were about current and future levels of progress. By contrast, the respondents from middle- and lower-income nations were both more knowledgeable about current trends and more optimistic about the future.

When asked, "In the last 20 years, the proportion of the world population living in extreme poverty has...?, all of the respondents from the wealthy nation choose "Increased" by a wide margin. In the most knowledgeable of the wealthy countries, Sweden, 39% believed that extreme poverty "Increased" and 30% believed that it "Decreased". All other respondents from wealthy countries chose Increased at least twice as much as Decreased. As we saw in Chapter One, the proportion of the world population living in extreme poverty has declined quite dramatically over the last 20 years.

Meanwhile, respondents from lower-income nations, such as Kenya, Senegal, India, Indonesia and Nigeria all had majorities who believed extreme poverty had decreased, usually by wide margins. Middle-income nations were much more widely dispersed in their answers.

Answers for the question "In the last 20 years, has the child mortality rate in developing countries increased, decreased or stayed about the same?" were more optimistic, but the same overall pattern emerges. Respondents from wealthy nations are less informed and substantially more negative than respondents from low-income nations.

Similar trends emerge in questions about developmental disabilities. Only on the question about trends in maternal mortality are wealthy nations solidly positive, although still less so than respondents from low-income nations.

When asked about future trends, the disparity between optimism from low-income nations and pessimism in wealthy nations intensifies. Interestingly, people are more positive about future trends in their community, less positive about future trends in their nations and least positive about future trends in the entire world. The further people get away from themselves, the more negative their answers. We will come back to these phenomena later.

When asked six different questions about trends over the next 15 years for the poorest countries, respondents from wealthy nations were overwhelmingly negative on all but one (gender equality). Again, respondents from low-income nations were overwhelmingly positive.

> **KEY INSIGHTS**
>
> Citizens of wealthy nations are ignorant of world progress, and because of that, pessimistic about the future. The rest of the world gets it.

While it is very good news that respondents from low-income nations are keenly aware of the progress over the last few decades and optimistic that it will continue, the responses from wealthy nations are disturbing. Quite simply, people in wealthy nations are ignorant of world progress in most domains, and their negative attitudes towards current trends will make progress more difficult in the future.

There is strong reason to believe that a large proportion of citizens believe that something is fundamentally wrong with our society. Whereas for decades major political parties in wealthy nations tended to be either center-left or center-right, the political center is rapidly shrinking. Political parties on both the left and the right are moving from the center towards the extremes. They each preach that things are bad, and they are getting worse. Both find receptive audiences.

So why is this? Why are people in wealthy nations so ignorant of progress and so pessimistic about the future? Why doesn't their prosperity make them more aware and more hopeful for the future?

I do not believe that there is only one answer to that question. Instead, I believe that there are a host of psychological, political and institutional reasons for this ignorance and pessimism.

It Is Not About the Facts

Could it be that people are just not aware of the facts?

Ignorance is part of the problem, but it is important to understand that it is intentional ignorance. Facts that back up the concept of progress are simply too easy to find.

We should start with the fact that almost no one starts out with an open mind and then decides to do a research project to figure out whether progress exists or not. (Just so we are clear, I did not arrive at the idea of progress before writing my book. I originally thought that this would be a history book, but then, after five years of research, I realized that "progress" was the only word for it).

At one time, it was hard to find good data about many of the positive trends that I documented in the first chapter. One had to purchase books at bookstores or go to the library. Not surprisingly, few people did so. Anyone with an intense interest in progress could get the facts, but those with just a passing interest had no easy way to get the facts.

With the internet and the rise of data-oriented websites and indexes, it is now very easy to get the facts. Perhaps the best means to do so today is to visit "Our World In Data" and browse their data for one hour. While that site bends over backward to present problems as well as long-term trends, it is hard to see how an objective person could come away from that site without a firm belief in progress.

Another hint of the problem is the way progress cynics act when confronted with the facts. In my experience, they do not update their thinking in any way. Maybe it is possible to create some doubt within the span of one conversation, but then when they interact with other progress cynics, they snap right back to their original thinking.

Unfortunately, for many people, the facts just do not matter. This chapter helps us to understand why.

Of course, this does not mean that supporters of progress should avoid fact-based arguments and quantitative data. Far from it. We should base our arguments on real data, and even more importantly, update our arguments as the facts change. But believers in progress must realize that we are battling against very powerful psychological forces, not ignorance of facts.

A Reminder Of What Progress Is Not About

I mentioned this briefly in the Introduction, but I want to emphasize what progress is not about to make my case clear.

The definition of progress that I use in this book series is "the sustained improvement in the material standard of living of a large group of people over a long period of time." In particular, I focus on changes to living standards that are rapid enough and sustained enough that one person could notice positive changes within their lifetime.

Progress is **not** about:

- The United States and Western Europe (it is about the entire world)
- The future (it is about the present and the past)
- What happened today, this week, this month or this year (one needs to focus on decades and centuries)
- Every single nation, sub-national group and individual enjoying the benefits of progress (there are always exceptions)
- The environment (it is about humans)
- Inequality
- A lack of bad events
- A lack of problems
- Utopia
- Consumerism
- Happiness (although I have given strong evidence that progress does lead to greater happiness)

Progress cynics joyfully point out the exceptions to progress, while ignoring the overall trend. I have no problem pointing out the exceptions to the rule. These are problems that we need to solve.

But relentlessly focusing on the exceptions, while deliberately obfuscating on the overall trends is dishonest and socially destructive. Progress cynics want to establish an anti-progress narrative that is not disrupted by reality. That is unacceptable.

Sometimes progress cynics lower their guard and admit that they do not disagree with the facts, but claim that they are only concerned about how those facts will be used by their political enemies. Statements like this, which I have frequently heard, show that they know that the facts cannot sustain their political narrative.

Progress cynics claim that they are worried about the concept caus-ing complacency about current problems. They conveniently ignore the fact that their cynical message is causing fear, despair and pessimism.

I understand that most people do not care much about progress or politics. Most people simply repeat what they hear other people who they trust say. Most people who are skeptical of progress are simply misinformed, and they can potentially change their minds.

My problem is with the public intellectuals and institutions who actively promote progress cynicism. They do not care about the facts and are actively trying to disrupt the communication of those facts because it conflicts with their personal, institutional or ideological self-interest.

Cynicism in progress has effectively become an identity badge with the Left becoming increasingly hostile to progress and the Right becoming increasingly skeptical of progress. To simplify somewhat, the Left sees the concept of progress as a threat to their narratives, which are designed to use the power of government to transform societies. The Right sees the concept of progress as eating away at the founda-tions of everything that they hold dear.

Both sides made up their minds before looking at the facts, refuse to update their beliefs and argue against facts because they see belief in progress as undermining their ideological agendas. Unfortunately, there is no pro-progress Center with a counter-narrative based upon reality and facts. I hope my books play some role in changing that.

So what enables the public to unthinkingly repeat what these prog-ress cynics say?

Negativity Bias

As Steven Pinker argues in *Enlightenment Now*, perhaps the biggest reason why people ignore progress and are pessimistic about the future is that we are born with a negativity bias. Instinctively, humans react more strongly to negative information than positive information.

From an evolutionary perspective, this makes perfect sense. Our brains are designed to enable us to survive, not be happy. Imagine you are a Hunger-Gatherer roaming the African savannah. You see a cluster of bushes and immediately two thoughts pop into your head. The first is "There are often lions near a cluster of bushes". The second thought is "There is often food near a cluster of bushes".

KEY INSIGHTS

Skepticism of progress is not due to an ignorance of the facts. It is due to cognitive biases in the human brain.

Any person who does not take that first impulse seriously might suffer a sudden death, while a person who does not take seriously the second might pass up one meal. The difference in consequences are just too great. Now if the person is near starvation, the consequences might even up, but that person must still be extremely aware of the threat of lions.

This phenomenon is well known to psychologists. In psychological experiments involving financial transactions, humans react far more strongly to losing ten dollars than winning the same amount. In assessing the character of a person, participants place more weight on bad actions made by the person compared to good actions taken by that same person.

This causes people to view what is happening in the world very differently. No matter how good things are as a result of progress, there are always bad things going on: wars, diseases, recessions, unemployment, and poverty. The brain naturally weighs bad things much more heavily than good things. It is natural for people to believe "the world is going to hell" while being surrounded by widespread progress.

Another important psychological principle is that people tend to view other people outside their immediate personal contact as being worse off than they really are. People generally view the general public as financially worse off than they really are.

The flip side of this is the positivity bias, where people tend to have a more positive view of themselves than they really deserve. So people tend to overestimate themselves and underestimate the general public.

Since the concept of progress is mainly about the outside world, people tend to exaggerate the negative and believe that the progress that they feel in their own life is due to their own accomplishments. "I am doing better off today than ten years ago because I work hard, but those other people who don't work as hard as I do are having it real bad."

Another part of the negativity bias is that people are far more likely to believe negative predictions of the future than positive predictions of the future. Anyone who makes a negative prediction of the future is viewed as wise and intelligent, while those who make positive predictions are viewed as uninformed wishful thinkers.

This bias tends to cause pessimistic experts to get far more attention than optimistic experts. The media has learned this, so they are more likely to cover pessimistic experts than optimistic experts. To advance their careers, experts gradually learn to tailor their message to get more attention, so it becomes a self-fulfilling prophecy.

Optimists are gradually squeezed out of the public debate by pessimists. This creates the false impression that all smart people realize that everything is bad and that progress does not exist.

Unreasonable Standards for Comparison

A common phenomenon that I have noticed among progress cynics is that they compare the world as it is (or their country or their community) to an imagined ideal. This is particularly common on the ideological Left and among young people. No matter how much progress has been achieved, the world will never be able to compare favorably with an ideal state. Ideal states have no problems, no trade-offs, no people making choices.

At the same time, those on the ideological Right and older people typically compare society to nostalgic but distorted memories of how they think life used to be. Many of them see any departure from this beautiful vision that exist in their head as decadence or moral decline. They fail to see life, even their own life, as it really was.

The proper standard of comparison is not with ideal images but with how people actually lived in the past or with other societies at the same time period. This is a much fairer standard by which to judge. In previous chapters, I gave a great deal of evidence that there has been progress across the world on many different development metrics.

Despite all this progress, however, no nation today can match the purity of an imagined society. Unfortunately, many people today seem to use an imagined society as their standard, and this standard makes all progress seemingly disappear.

Progress is Ignored in Schools

While understanding past and current progress is critical to maintaining future progress, our educational system does not teach it. Most content in school has nothing to do with progress, for example, reading, writing, mathematics, sciences, and foreign languages. The amount of time that students spend studying history is a relatively small portion of the school year. And when students learn history, it is mainly about names, dates and events.

There is almost no class time devoted to long-term trends of progress in everyday life (although I think this would make a much more interesting topic for students). Students typically graduate secondary school and even university with almost no grasp of how much progress in the past has affected their lives.

Generational Reset

I believe one of the main reasons why people in wealthy nations are so ignorant of progress is what I call "generational reset." Because few people study history, each generation has very little knowledge of the progress that took place before they were born. Even during their childhood and early adolescence, most people are not very aware of the world outside their immediate family and friends. For most people, the only progress that matters is the progress that has taken place since their late teens.

For a person who was born in 1990, all progress before that year is irrelevant. Even the progress that took place in the 1990s is a distant memory. So for a young person born in 1990, they are probably only aware of any progress that has taken place between say 2008 and 2020. In such a short time span, it is a very high bar to show any substantial progress.

Similarly, for a person born in 1940, all the progress that took place before their adolescence is irrelevant, consigned to the dustbin of history. The enormous progress that happened in Western countries in the preceding century is perhaps sensed, but it is not fully integrated into their perceptions of the world.

At the same time, the bottom limit for what is acceptable is also set early in their lifetime. Anything that gets worse than what occurred previously, no matter how small, is perceived strongly. This means that any bad thing that involves a loss is felt much more strongly than many other good things that occur.

For this reason, a single bad event or trend, a terrorist attack, war, epidemic, recession, is treated as powerful evidence that things are getting worse. All the periods of peace and prosperity have far less of an enduring effect on our psyche.

Progress Is Smooth, but Problems Are Spiky

As one can see from the graphs in Chapter One, progress generally consists of long, slow trends that take time to have a noticeable effect.

Bad events, on the other hand — wars, violence, famines, epidemics and natural disasters come on fast and are very noticeable. Usually they disappear in a relatively brief period of time. Because there are almost always some bad things going on even during periods of great prosperity, it is easy to focus on the bad things.

This is one reason that I believe studying long periods of history is a serious antidote to cynicism. Doing so forces one to ignore sudden changes that come and go and, instead, to focus on long-term trends. It is hard to miss the long-term trends of progress if you look at a long enough timescale.

Negative Media Coverage

As Steven Pinker points out, another factor undermining people's perception of progress is negative media coverage. The primary means by which people learn about what is going on in the outside world is the media. While it might be via radio, television, newspapers or the internet, the media has some common themes that shape their coverage.

Media researchers, such as Darrell West, have long known that media coverage is overwhelmingly biased towards the negative, the short-term and strong emotional connection. There are an enormous number of studies that show that media coverage is overwhelmingly biased to cover negative events over positive events.

KEY INSIGHTS

The media, politicians, interest groups, social media and ideologues have a strong self-interest in you not seeing the progress that surrounds you. Negativity and cynicism give them viewers, voters and money.

While many journalists feel that they have an obligation to inform the public, the overriding goal of news organizations is to generate high ratings. It is much easier to get people's attention by shocking them rather than by pleasing them. It is only really in the domains of economic and sporting news that positive events get high levels of coverage. News about progress does not fit into the negativity bias that dominates the media.

In addition, the media has a strong recency bias. The media overwhelmingly covers what happened today, yesterday and this week. Unfortunately, progress is overwhelmingly about long-term trends that show little day-to-day variation.

It is simply not considered news that infant mortality declined 0.0001 percent today. While the media occasionally covers long-term trends, usually in the form of annual reviews in late December, it is not a very large part of the coverage. For that reason, most journalists simply do not consider trends of progress to be news.

News organizations also seek to create an emotional connection with their viewers. The easiest way to do so is to show strong action-oriented visual images of negative events that just happened: a fire, murder, storm, coup, war or natural disaster. Long-term trends simply do not lend themselves to creating this emotional connection in the same way.

When you combine all of these factors, the media presents a highly distorted view of the world. It depicts a world without progress where seemingly random bad events hurt helpless victims. The real lives that normal people lead and the progress that they experience are almost completely absent.

Politicians and Interest Groups

People involved in politics, whether they are political activists, candidates or leaders of interest groups have a strong incentive to focus on the negative. Any interest group that claims that things are going well will find it hard to motivate donors to give them money. Political activists that do the same will find it hard to motivate people to protest or vote. Political candidates who are running against incumbents have a huge incentive to focus heavily on the negative to get votes. In addition, incumbents outside the governing coalition also have a strong incentive to focus on the negative.

The only people in politics who have an incentive to focus on the positive are incumbents of the majority party running for reelection. They have the incentive to say that the economy is doing well as well as talk about all their wonderful legislative accomplishments. But for every incumbent, there is at least one challenger who has a very large incentive to focus on the negative.

Regardless of whether they are an incumbent or challenger, candidates often engage in negative campaigns that accentuate how bad their opponent is. While there is evidence that this hurts the person running the negative campaign, it is clear that it hurts their opponent even more.

And even incumbent candidates that run positive campaigns do not claim that there has been progress in society. They merely claim that their leadership and legislative accomplishments have made things better. Any candidate who claims that there has been a great deal of progress and they had nothing to do with it would probably have a short political career.

No one in the political world makes the claim advanced by this book: that there has been a great deal of progress and not much of it was caused by short-term changes in government policy.

Relentless Negativity of Political Ideologues

Political ideologues of both the right and left have an enormous incentive to focus on the negative. One thing that virtually all ideologues agree on is that "Things are bad, and they are getting worse."

If people believe that conditions are the best that they have ever been and this trend is likely to persist in the future, ideologues will garner very little support. Ideologues need to persuade people that things are worse than they actually are.

Only by convincing people that the system is not working can political ideologues convince voters that radical changes are necessary. So they relentlessly propagandize about how terrible things are.

In fairness, ideologues often identify very real problems. The problem is that they fail to put them into the proper context, and they do so on purpose. The world is full of millions of little problems that are gradually being solved, usually by the richest nations. Poorer nations then copy those solutions and modify them to fit the local environment.

Some problems are already diminishing in scope and just continuing on our present course is the best solution. Some problems are not solvable under current technologies. Some supposed "solutions" create new problems that are worse than the original problem. Some solutions use up valuable monetary and human resources that would be better used for solving other more serious problems. Most importantly, the vast decentralized problem-solving network that is a modern society is usually far better at solving problems than the solutions presented by ideologues.

Even worse, political ideologues fail to propose workable solutions. One thing that is constant in many political movements, particularly those that protest in the streets or use violence is that they focus far more on the problem than the solution. They want us to "wake up", as if society is full of sleepwalkers who are doing nothing. But while the ideologues are attacking society, society at large is going about the real task of identifying solutions, testing those solutions, modifying them based upon results and then scaling them up so all of society benefits.

Identifying problems is very easy. Determining which problems are worth focusing resources on is much tougher. Coming up with possible solutions is also relatively easy. But it is much harder to implement them effectively, test their results and then either modify them or toss them out based upon results. Most problems are already being worked on, with varying degrees of success.

> **KEY INSIGHTS**
> Most so-called crises are actually just one of the thousands of solvable problems. Let's let the decentralized network solve them.

Overuse of the Word "Crisis"

Perhaps the most overused term in politics is the word "crisis." While policy experts, leaders and activists all play important roles in bringing problems to public attention, they generally go far beyond that. They know that the government is simultaneously trying to deal with thousands of different problems. If policy experts, leaders and activists claim their domain has a problem, then they are just adding one more to a long list. The chances of immediate government action are close to zero.

When policy experts, leaders and activists claim that their domain is experiencing a crisis, however, their favored issue can potentially jump to the top of the agenda. By claiming that something is a crisis rather than a problem, they attempt to promote their favorite issue to a higher rank than all the other problems competing to get on the political agenda.

In reality, the world faces thousands if not millions of problems. Society can be thought of as the ongoing actions of millions of people solving problems, most of which have been known for decades or centuries.

But a crisis is so severe and so urgent that everything else should be subordinated to solving that problem. And in a true crisis, there is no time to discuss the severity of the problem compared to other problems and possible alternative solutions. We must act now! (And "act" generally means whatever policy solution that one person happens to believe in). While some problems are serious enough to be considered a crisis, more often than not it is best to allow society to continue its natural problem-solving process.

Rise of the Ideological News Media

Over the past decade, there has been a big change in how the traditional American media cover the news. In the 1990s, news outlets such as CNN, ABC, NBC, CBS, Fox, the New York Times, the Washington Post, and the Los Angeles Times set the media agenda. Journalists and editors were overwhelmingly left-of-center, but they typically worked hard to present a balanced and "objective" view.

In practice, this was more of a center-left consensus that pretended reasonably well to be objective, but at least it was not too far from the center of American politics. Journalists and editors believed in at least appearing to be objective, partly out of professional integrity, but also because they were dependent upon ad revenue from largely conservative corporations. An uneasy truce existed between the goals of liberal reporters and the goals of conservative corporations.

Not that all the media bought into this consensus. There were plenty of ideological outlets, but few of them focused on news. Magazines such as the Nation, the New Republic, the Atlantic, Mother Jones, National Review, The Economist, and Reason all had explicit ideological slants. While their content focused on big issues of the day, their monthly publishing cycle forced them to think more long-term than the daily news cycle of the news media.

It is also important to realize that these overtly ideological outlets were not directly tied to either the Democratic or Republican party. Writers typically chose partisan sides, but they felt an obligation to offer constructive criticism of their own side. While writers certainly felt pressure to toe the ideological line, they were not under intense pressure to toe the partisan line.

The internet gradually undermined the old business model that "objective" news reporting was based upon. Local newspapers, in particular, saw their paper subscribers decline rapidly. Some were able to keep up their ratings by publishing digital versions of their newspaper. Then the rise of Craigslist seriously undermined their flow of revenue from want ads, causing a deep decline in revenue.

First, the Fox News channel broke off from the "objective" news consensus. They pointed out that this consensus was not objective, but was actually center-left. By stealing conservative viewers from the networks, they increased the incentive of those networks to move to the left. As traditional news organizations saw their entire business model collapsing, they frantically searched for financially-viable alternatives.

Starting around 2010, a new business model for news came into being. Focus groups among elite news organizations concluded that overtly ideological news coverage that reinforced viewers existing ideological perspectives was the most profitable business model. This new business model effectively converted traditional news organizations into ideological outlets that used the news to create an ideological narrative.

The new media shifted from covering fires, murders, storms, coups, wars and natural disasters to partisan combat and grand ideological narratives. Now the goal was not to inform or even to shock to boost ratings. The goal was to create anger and resentment against the opposing party, their ideology and their followers.

So now the self-interest and organizational interest of political parties, politicians, interest groups and the media had effectively been fused. While this fusion is based upon ideology, it can best be understood as a business model based upon self-interest. To get votes, money and viewers, all these institutions knew that they had to keep the focus on the idea that "Things are bad, and they are getting worse."

The business model based upon an ideological narrative was the opposite of the reality of progress. So an increasing portion of American institutions has a strong incentive to ignore progress and propagandize for a much more cynical ideological agenda.

Politics is Increasingly about Moral Identity

The past few generations within wealthy nations have been born into unique circumstances. While previous generations had to struggle to survive, almost everyone in wealthy nations today lives in such good circumstances that survival is not at stake. Certainly, some are much better off than others, but very few people are constantly in a situation where they might die soon. Prosperity is no longer something that one needs to struggle to achieve. It is now a fact of life.

I believe that this has affected how people think about politics. Politics is no longer about group survival. Now it is an expression of moral identity. People increasingly support political parties not because the party is going to help them and people like themselves survive, but because they share a common moral vision. And people typically support policies, not because they work, but because it aligns with their moral identity.

To make things worse, many people see ignoring results as a badge of personal morality, not something that reflects a personal flaw that needs correcting. They do not recognize that ignoring results means ignoring people. Forced to choose between their ideals and the results of implementing their ideals on real people, far too many people choose to selfishly hold onto their ideals regardless of results.

So if we put all of the factors listed above, we have a large majority of people in wealthy societies who are ignorant of progress, fearful of losing what they have, surrounded by the media and political actors who are constantly focused on how bad things are. No wonder so many people in wealthy countries are ignorant of progress and negative about the future.

Conclusion

Unfortunately, despite all the progress that surrounds us, pessimism and cynicism are widespread throughout the wealthiest nations. Rather than comparing our present state to other nations or societies of the past, many people appear to compare today with unachievable visions for how societies should be or nostalgia for how they imagine things used to be.

A relentless cynicism seems to permeate politics, the media, the educational system and an increasing number of institutions. Rather than presenting current problems in the context of solutions that have already worked, problems are identified as crises that threaten to overwhelm us unless we immediately make a radical change.

Few see modern society for what it is: a vast, decentralized problem-solving network that has already solved millions of problems and will continue to do so if we let it.

CURRENT CONSTRAINTS TO PROGRESS

There has been enormous progress throughout the world over the last three decades. Inequalities still exist, though. Most likely they will always do so. Fortunately, they are diminishing greatly over time, but true equality eludes us.

Large inequalities between nations and subnational groups are one of the key reasons why the Left is skeptical of or hostile to progress. Typically, the Left views prosperity and equality as the expected outcome and any deviation from those outcomes as something that needs to be explained.

The Left explains variations between societies and within societies as being caused by discrimination, selfishness, capitalism, and colonialism. More sophisticated explanations on the Left point to structural economic factors, politics or ignorance causing inequality.

Anyone who has got this far reading this book should be immediately skeptical of these explanations. We have seen that inequality has been rampant in all societies that evolved past Hunter-Gatherer societies. We have also seen that poverty has been an inherent part of the human experience for hundreds of thousands of years. We have also seen how differences in wealth between societies are very similar to those 2,000 years ago.

We have also seen that progress was invented in a few Commercial societies centuries ago and that progress diffused through the rest of the world over centuries. Progress and prosperity are unusual; poverty is the norm. I hope that my book helps you to better understand how progress emerged and gradually diffused.

We have also seen that six historical breakthroughs have led to sudden expansions of progress. The most recent was the fall of the Soviet Union in the early 1990s.

But why has it taken so long for progress to spread throughout the globe? Why didn't the emergence of Commercial societies rapidly spread? Why didn't the entire world industrialize within just a few decades of Britain doing so? Why are some societies still living in circumstances that more closely resemble their lives 2,000 years ago than what Western nations today enjoy?

Part of the reason why the Leftist explanations of discrimination, selfishness, capitalism, colonialism, structural economic factors, politics and ignorance have lived on despite being wrong is that there is no real counter-explanations. This book starts the process of creating that counter-explanation.

Major constraints on the diffusion of progress to poorer nations include:

- Geographical constraints on food production that limited the type and amount of food that society could produce. Those constraints on food production placed fundamental limits on which types of societies could evolve.
- Extractive institutions in Agrarian societies that diverted the food surplus towards increasing the income and status of elites who controlled those institutions.
- Geographical distance
- Cultural distance

- Monopolies
- The legacy of society types
- Ethnic/religious/racial identities
- Ideologies that promote class, ethnic, religious or racial resentments.

We have already discussed the first two constraints in previous chapters. Fortunately, both of these constraints, food production and extractive institutions, have been largely overcome in most of the wealthy societies of today.

In this chapter, I will focus on the constraints that still exist today.

Geographical Distance

Geographical distance has always been a barrier to the diffusion of progress. Before the advent of large ocean-going sailing ships, societies that were far away from each other rarely came into contact. We have already seen that the development of North America and South America was seriously constrained by a lack of contact with more complex Eurasian societies before the year 1500. The same can be said for Sub-Saharan Africa, New Guinea and Australia. It is likely that if these regions had been closer to Eurasia, particularly Northwest Europe, they would have developed far more complex societies.

Societies on the Eurasian continent were able to develop at a far more rapid pace because they came into far more frequent contact with each other. Innovations in one society could be noticed by traveling merchants and then brought back to their homeland. Just as important, because the Eurasian societies were in constant military competition, any laggard society would be forced to copy more complex technologies, skills and social organization or risk annihilation through conquest.

Fortunately, today we live in a world where transportation and communication technologies make the barrier of geographical distance easier to overcome than ever before. Most people on the planet now have relatively easy access to modern transportation and communication devices that were inaccessible to all but the wealthiest just a few generations ago. Mobile devices, in particular, have diffused across the world with amazing rapidity. In doing so, they have transformed the lives of so many previously poor people by connecting them.

But it is important to realize that geographical distance still plays an important role. Rural regions located far from trade-based cities often lack transportation and communication infrastructures. Inland regions of South America and Africa are often very remote from navigable rivers and container ports. Sub-Saharan Africa, in particular, lacks a robust electrical grid capable of powering modern communication devices reliably.

Cultural Distance

Another barrier to the diffusion of technologies, skills and social organizations is cultural distance. Cultural distance is a measure of similarity between people based upon the culture that they were born into. In general, cultural distance aligns with geographical distance, but due to large-scale migrations, they sometimes differ substantially.

For example, the English are culturally similar to their neighbors the Scots, but the English are more culturally similar to the distant Australians and New Zealanders than to their French neighbors. The reason, of course, was the mass immigration of the British to Australia and New Zealand.

But how does one measure cultural distance? One method is to list all the characteristics of every culture and compare them to each other, but this would be extremely time-consuming and very subjective. Inevitably, researchers would disagree as to which characteristics to include and how important each of them is compared to the other.

Genetic sequencing technology gives us a simpler and more objective method: genetic distance. By measuring the aggregated differences in allele frequency on a chromosome, geneticists can measure how similar two peoples are to each other genetically.

In general, the overall amount of difference between two peoples is related to migrations in the distant past. The key assumption is that when two peoples are relatively isolated from each other, they will diverge both genetically and culturally. So the longer that two peoples have lived since their last common ancestor, the more different their cultures are likely to be.

> **KEY INSIGHTS**
>
> Societies tend to copy other societies that are close to them geographically and culturally.

It is important to point out that this method does not assume that genes create a culture and that culture is immutable. Genetic sequencing is merely a convenient method to measure cultural distance, and there is strong reason to believe that genetic distance and cultural distance are closely related.

In an article entitled "The Diffusion of Development", Enrico Spolaore and Romain Wacziarg tested how much of an impact cultural distance had on the ability of poor countries to copy wealthier countries both today and in previous centuries. Not surprisingly, they found that geographical distance and cultural distance are closely related. In other words, nations close to each other tend to be culturally similar.

Far more interestingly, they find that cultural distance has a more powerful effect than geographical distance and that nearby nations with significantly different cultures have particularly strong barriers to diffusion. These results help to explain why industrial technologies quickly spread from Britain to the United States and Canada (i.e. between geographically distant but culturally close nations), but much more slowly from Southern Europe to North Africa (i.e. between geographically close but culturally distant nations).

Spolaore and Wacziarg find that after controlling for other possible explanations, cultural distance accounted for 49% of the variation in income differences between 1500 and 1820. Cultural differences played a crucial role in hindering the diffusion of innovation. In 1870, the peak of the Industrial Revolution, the effect increased to 80%. Cultures that were similar to the cultures of the first industrializers were far more willing to copy industrial technologies than those that were more distant culturally.

Fortunately, in more recent years, the barriers of cultural distance have declined (to just under 20%) as a few non-Western nations copied industrial technologies and these could be in turn copied by other nations that were culturally similar. In East Asia Japan played a particularly important role in this regard.

Note that cultural distance is not just a barrier to the diffusion of progress; it is also a local accelerator. If an innovation takes place among one's group, members of that group are much more likely to adopt that innovation. People tend to copy other people who are like themselves. So culture accelerates diffusion within cultural groups. But given that humanity is divided up into hundreds or thousands of cultural groups, this diffusion within cultures can take us only so far. Overall, culture tends to retard diffusion.

Monopolies

Earlier we saw that organizations competing against each other is one of the key driving forces in progress. Because there are limited resources, usually in the form of revenue, organizations are forced to compete against each other for survival. This competition forces each organization to adopt new technologies, hire people with skills in those technologies and adopt new processes to most effectively coordinate these technologies and skills. In this way, organizations that compete against each other facilitate progress.

There are, however, some organizations that do not need to compete. Most commonly these are governments or government-sanctioned private monopolies. Because these types of organizations are insulated from competition, they have little desire to adopt new technologies, skills or processes. This does not stop progress, but it does slow it down considerably. The higher the proportion of the total economy that is controlled by monopolies, the slower the progress.

Most organizations would prefer to be a monopoly. Being a monopoly creates a steady stream of revenue without the need for unsettling change. Monopolies create a comfortable internal environment for the members of that organization. No one has to improve or change in any way to maintain a steady flow of revenue.

The winners of a monopoly are its members, particularly those at the top who control the organization. The losers are the rest of society.

Because the benefits of being a monopoly are so great, most organizations try to become one. Businesses push governments to enact regulations, tariffs, subsidies or loopholes to gain an advantage over their competition. Businesses often try to buy start-ups that could potentially grow into dangerous competitors. They also narrow their product line so that they can gain a significantly larger portion of a smaller market, enabling them to have more influence over prices. Businesses sometimes form cartels to fix prices.

> **KEY INSIGHTS**
>
> Monopolies tend to undermine progress because they have a lower incentive to innovate or copy successful innovations.

Governments, in particular, want to preserve their monopoly. Those who work in government are naturally suspicious of private organizations that can potentially out-compete them. There has also been a long-term trend within the wealthiest nations for centralized governments to grow at the expense of local governments. While local governments are often forced to compete against each other, centralized governments can often avoid this. There has also been a long-term trend of international organizations growing at the expense of national governments.

All of the above suggests that we should be deeply suspicious of monopolies. While it is likely that some monopolies are inevitable, most are not. Limiting the number and size by promoting transparency and competition should be one of the key goals of government policy.

Of particular importance is enabling new organizations to start up and old uncompetitive organizations to fail. As long as government policies make it difficult for new organizations to be founded, it will be much easier for existing organizations to maintain their power. In addition, if government policies try to maintain existing organizations even when they are not delivering benefits to society, this will tend to stifle progress.

The Legacy of Society Types

In this section, I want to push the explanation for inequalities between societies into areas where we do not yet have enough data to make substantive conclusions. I openly admit that what follows in this section is speculative, but it fits in well with what this book has already established.

What I call the "legacy of society types" probably plays a big role in explaining why progress diffused to some societies much faster than others. This concept combines an explanation as to why the rank-order of national wealth has been so stable over the last 2,000 years with the fact that some nations have been able to transform themselves within one generation (i.e. the Transformative 16 and the Wealthy 12 that we discussed in the first chapter).

It is striking that even the nations of East Asia that have experienced such explosive economic growth — Japan, South Korea, Hong Kong, Taiwan and Singapore — have still not reached the same per capita GDP as the United States. But as more and more nations accelerate in their copying of technology, skills and social organizations from wealthier nations, the differences are narrowing quickly.

I am fascinated by the fact that the current standard of living of a nation, including ethnic minorities within nations as well as the timing of industrialization are closely tied to the society type of their genetic ancestors in 1500. For example, the Dutch people today are the genetic descendants of people who lived in Commercial societies in 1500. The Japanese people today are the genetic descendants of people who lived in Agrarian societies in 1500. The Sub-Saharan African people today are the genetic descendants of people who lived in Horticultural and Herding societies in 1500.

Those simple facts tell us a great deal about their current standard of living, the timing of their transition into Industrial societies and their relative outcome on the development metrics in Chapter one. This is a subject I will return to in a future book.

Why Race Does Not Matter Very Much

Let me be clear, the argument that I am making has nothing to do with race. Race is a genetic characteristic of a person, but there is not the slightest evidence that it has an important impact on how societies in different regions evolve. The actual cause is geographic constraints on food production and the resulting type of societies that can develop.

Because both race and society types vary by region, this creates the illusion that race is important. It is not. This fact should be obvious to anyone who has read this far, but race has become such a controversial topic that it needs to be stated clearly and boldly.

Multiple society types have inhabited every continent and racial group. Regardless of the racial characteristics of their inhabitants, Hunter-Gatherer societies across the globe strongly resembled each other. As did Horticultural societies, Herding societies and Agrarian societies. Each region has certain geographical constraints that enabled certain society types to evolve and others to not do so. The race of the inhabitants was irrelevant.

There is not the slightest evidence that racial characteristics in any way change society types or prevent any from evolving to more complex types. If Europeans had evolved in geography similar to Sub-Saharan Africa, they would have evolved societies very similar to the ones that actual Africans lived in. And if Africans had evolved in geography similar to Europe, they would have evolved societies very similar to the ones that Europeans lived in.

People who are genetically European have lived within all seven society types. They started as Hunter-Gatherers and gradually evolved up the evolutionary sequence to Agrarian societies. They did so because of geographical factors that had nothing to do with race. Some of those nations evolved into Commercial societies because of geographical and political factors. Then all of Europe evolved into Industrial societies. Again this had nothing to do with the fact that they were genetically European.

People who are genetically Asian have also lived within all seven society types. Because of geographical factors that vary by region, different societies followed different evolutionary sequences. Some Asians evolved into Herding societies, some evolved into Horticultural societies, and some evolved into Agrarian societies. Most recently, many Asian societies evolved into Industrial societies. One race evolved into many different society types.

> **KEY INSIGHTS**
>
> It is geography and the resulting legacy of society types that explain modern inequality between societies, not race or racism.

People who are genetically African have lived within four different society types. Because of geographical factors that varied by region, some Africans evolved into Herding societies, some evolved into Horticultural societies, and some remained as Hunter-Gatherers. And many people today of African descent are living in prosperous Industrial societies and enjoying the benefits of doing so. The same is true of Native Americans and other indigenous peoples throughout the world.

Because both race and society types are related to geography, it is easy for people to mix the two up. Because the race of a person is immediately visible to another person, it is easy to assume that characteristics or behaviors are driven by race, rather than the legacy of their ancestors' society type.

It is not an accident that radical ideologies on both the left and right obsess over the concept of race. Ideologies always obsess over certain demographic factors to pit them in conflict with each other. One is labeled the oppressor and the other is labeled the oppressed.

Ideologues conveniently paint themselves as rescuers of the oppressed, who are said to not fully understand their self-interest. The ideologues claim that if they are given total power, they will liberate the oppressed and bring progress to them.

It should be clear, though, that radical ideologues sabotage progress and individual success rather than promoting it. Progress comes out of the natural working of the vast, decentralized problem-solving network that is a modern society. The "solutions" proposed by ideologues attack the very foundations of progress.

Those who promote racial identities undermine progress. When one believes that some people become successful by exploiting the rest of society, this undermines the moral legitimacy of the successful. This undermines the desire of people to copy the successful.

When people believe that their group is inherently more moral than other groups, this undermines the desire of members of that group to copy successful people in other groups. The strong centralized institutions promoted by ideologues undermine the diversity and experimentation that is necessary for success.

Integration, intermarriage and copying the successful, even if they are from other racial groups, is the way to promote success for racial minorities. Radical ideologies and racial identities will only undermine their success.

I hope that the concepts in this book enable us to view the world in a non-racial way. I hope that if people start recognizing the importance of geographic constraints, food production and the radically varying amounts of innovation within differing society types, we can move past the extremely unproductive discussions that revolve around the concept of race.

Progress is an evolutionary process that has been gradually spreading across the planet for 800 years. It has been a great force for positive change in people's lives. Because progress is the outcome of millions of decisions made by regular people, each of us can either promote or hinder progress. In particular, we can make decisions that enable us to enjoy the benefits of the progress that surrounds us.

Ethnic/Religious/Racial Identities

Earlier we learned that cultural distance creates barriers to the diffusion of innovations in technologies, skills and social organizations. So why is cultural distance such a barrier? Why don't people objectively compare results and copy those that seem to work best?

Paul Richerson and Robert Boyd's theory of Cultural Evolution helps us to answer this question. They posit that as human societies evolve, groups naturally tend to differentiate. People start to adopt the customs and values of other people near themselves.

At first, this might seem to predict long cultural unification, but the actual results are the opposite. As people who live nearby become more similar, they also become more different from people in distant regions.

As cultures evolved humans created cultural markers to visually denote who belongs to their group. They evolved common languages, religions, dress, hairstyles and mannerisms to show fellow members that they are both in the same group.

Humans also tend to invest moral judgments about those differences: my group's markers are not just different from other groups, they are better. For a person who rarely comes into contact with other cultures, seemingly strange customs become not only normal but also desirable.

Because technologies, skills and social organizations are an important part of a culture, humans naturally invest them with moral worth. They are not just useful technologies, skills and social organizations; they are how we do things. They are part of what makes our culture meaningful. They may even be the will of the gods.

When we invest a moral quality to cultural identity, it is easy to see why culturally distant groups would resist copying technologies, skills and social organizations even if they were superior.

Those ways are not our ways. We have our own ways and they have worked perfectly well for as long as our people have been around. Why should we copy people who are so clearly strange and morally inferior?

While this way of thinking is declining in the 21st Century, ethnic/religious/racial identities are still a powerful barrier to poor groups copying the technologies, skills and social organizations of more successful groups. Elites in poor societies, in particular, have a vested interest in stopping diffusion. These new technologies and skills promote new social organizations that potentially fund political rivals. Elites in poor societies would rather be leaders of poor societies than middling members of prosperous societies.

For this reason, elites in poor societies try to mobilize fellow members of their ethno-religious-racial group against foreigners and other successful groups within their country. By harnessing the natural human need to be part of a cohesive moral community, elites can undermine the threat to their power that modernity represents.

While their people will benefit from copying more successful nations and minorities, most elites in poor countries do everything that they can to promote resentment towards the successful. They promote a zero-sum mentality where the success of someone else hurts us as a people. They promote the idea that those who are successful are hurting other people by being successful. They promote the idea that the group must stick together to fight back against their oppressors.

> **KEY INSIGHTS**
>
> Even the poorest groups can become wealthy by copying wealthier societies, but ethnic/religious/racial identities undermine their desire to do so. Leaders of poor nations and poor minorities have an incentive to promote resentment to preserve their power.

Promoting ethnic/religious/racial resentments against the successful has also become far more common in wealthy societies. While in 1970 most wealthy nations were quite homogeneous in all three dimensions, widespread immigration since that time has made almost all wealthy nations far more diverse.

Left-wing parties have increasingly focused their energies on promoting resentments within the descendants of immigrants against the more successful native population. Right-wing parties have increasingly focused their energies on promoting counter-resentments against immigrant populations. Both undermine the desire of ethnic/religious/racial minorities to copy the successful and integrate into the larger population.

There is nothing inherently bad about ethnic, religious or racial identities. Humans are social animals. We naturally identify ourselves with a group and feel a need to conform to the values and behaviors of that group. This is very unlikely to change, no matter how much progress the world experiences.

Ethnic, religious and racial identities can also be harnessed as a means of promoting the idea that poor people should copy the technologies, skills and social organizations of more successful individuals and peoples. Political leaders can challenge their followers to acknowledge problems and devote resources to catching up.

Unfortunately, political leaders far more often promote resentment of the successful. They believe this will help their people, when in fact this resentment keeps their people from enjoying the benefits of progress. In future books in this series, I hope to be able to cover this topic more thoroughly.

The Threat of Radical Ideologies

Most of the constraints on progress go back millennia. In 1917, however, a new threat to progress emerged: totalitarian ideologies. Just when repressive Agrarian societies in Europe were collapsing under the pressure of World War I and amid Woodrow Wilson's dreams of worldwide democracy, freedom and trade looked like they might be realized, Vladimir Lenin and the Russian Bolsheviks came to power.

The Bolsheviks played on class resentments against more successful groups in society. They used this resentment to establish extremely centralized regimes that subordinated all decisions in society to Communist doctrine. The Communists quickly gained control over all political, military, economic, religious and cultural organizations. Nothing was outside the purview of the state and ideology, not even the family.

The Bolsheviks created a template for action that dominated much of the 20th Century: a small dedicated group of ideologues achieving power through violence. Once in power, those ideologues subordinated everything to the centralized state. All alternate views and ways of doing things were to be stamped out through indoctrination and violence. A viewpoint less conducive to bottom-up progress is hard to imagine.

From 1917 until 1991 the radical ideologies of Fascism, Nazism, Communism and Authoritarian Socialism did enormous damage to humanity. These ideologies focused class, ethnic, religious and racial resentment upon more successful groups in their own society and abroad.

These regimes copied the technologies innovated by much freer Industrial societies and transformed them from agents of progress that benefited the masses into agents of death and rigid conformity. Ideologues started wars, committed genocides and established some of the most brutal regimes in world history. While World War II stamped out the threats of Nazism and Fascism, that same war also gave ideologies of the Left a new burst of life.

KEY INSIGHTS

The biggest threats to progress are radical ideologies that deliberately intensify ethnic/religious/racial resentments against more successful groups. This undermines the desire of the poor to copy the rich (their only pathway to progress).

Since 1917, progress has been under threat by ideologies on both the right and left that seek to radically centralize political power and subordinate all decisions within society to ideology. While the old Agrarian elites sought to use centralized institutions to extract wealth from society, new ideological movements seek to use the same methods to transform society towards utopian ideals.

The final collapse of the Soviet Union in 1991 unleashed the greatest wave of progress that the world has ever known. Nation after nation abandoned Communism and Authoritarian Socialism in favor of allowing people to solve their own problems. Faith in the ability of the all-powerful centralized state to deliver utopian solutions collapsed.

Before 1991, wealthy Western nations and a few East Asian countries enjoyed the bulk of the progress. After that date, the vast majority of mankind has done so. As we saw in previous chapters, that progress included major achievements in per capita GDP, education, health, sanitation, water supply, neonatal mortality, longevity and declining levels of violence.

Unfortunately, the threat from ideologies has not completely disappeared. The threat has merely been transformed. In both Europe and the United States, there has been a tremendous polarization of politics. Whereas one large center-left party and one large center-right party typically dominated politics in 1990, most democratic nations have recently experienced a surge of left-wing and right-wing politics.

In particular, college-educated professionals have moved from supporting center-right parties to left-wing politics. From the late 1940s until the 1960s, members of the working class who wanted to use the power of government to provide economic growth and security dominated the voters for parties of the left. For this reason, parties of the left generally favored more spending on pensions, health care, education and unemployment benefits. But the center-left leaders also knew that a market-based economy was essential for paying for those programs.

Starting in the 1960s and gradually building over the decades, college-educated professionals joined the parties of the center-left. They gradually transformed party ideology to focus on the environment, peace, gender, race and immigration. All of these issues were ones where the traditional working-class constituencies of the center-left parties took stands that were closer to the parties on the right. Pro-immigration policies by center-left parties, in particular, provoked a strong backlash.

The result since 1990 has been a rapid decline in center-left parties throughout Europe and a rapid increase of ideological parties on both the left and the right. In this polarized environment, college-educated professionals have moved rapidly to the left and working-class voters have moved rapidly to the right.

Both sides are hostile to the idea of progress. They both subscribe to the doctrine of "Things are bad, and they are getting worse." And they both strive to control the government and implement policies that will likely make these statements come true.

Politics is no longer about making incremental changes to redistribute the gains of progress. It is now a zero-sum ideological combat that focuses on identity and resentment.

Throughout human history, religion was the principal means by which humans determined what is moral and what is immoral. Religious texts and traditions were passed down from generation to generation. Even as late as 1900, virtually all people believed in some sort of religion that structured their lives and how they should act towards themselves and others.

With the extraordinary growth of scientific knowledge since the 1840s, an increasing portion of educated people in Western societies has abandoned religion as the key shaper of their morality. Many people hoped and believed that religion was being replaced by logic, science and reason.

While it is true that we have a far greater understanding of the natural world than we used to, I do not believe that religion is being replaced by logic, science and reason. Except for sociopaths, humans hunger for morality and desperately want to believe that they are moral beings. Perhaps more importantly, they want other members of society to perceive them as moral people. Science and reason can fill our brains with facts, but they cannot convince us that we are moral beings.

Ideology, particularly left-wing ideology, is increasingly replacing religion among educated people in Western nations. In their worldview, it is the ideology that increasingly dictates what is moral and immoral. And ideology pushes people to use the power of government to create a more just society, even if it means compelling others who do not share that vision to comply.

While the substitution of ideology for religion is not inherently bad, there are reasons for concern. Most religions, with Islam being a clear exception, create a clear distinction between religious beliefs and government authority. Religions primarily focus on the afterlife, the unknown, morality and how individuals should treat each other. They generally say very little about what governments or other institutions should do.

Ideologies primarily focus on what governments should do and what individuals should do to change government policy. This inevitably leads those who uphold an ideology to want to centralize government power and then use that power for implementing the dictates of the ideology.

Because progress primarily comes from society self-organizing, experimenting with solutions and then copying the solutions that work, a centralized government driven by ideology inherently attacks the very foundations of progress. While the great enemies of progress in the past were centralized extractive institutions designed to satisfy the *material* interests of elites, today the enemy of progress is extractive institutions designed to satisfy the *ideological* interests of the affluent.

A particular problem is that ideological leaders deliberately cultivate resentments against successful people of different ethnicities, religions or races. These leaders seek power by intensifying the natural human bias towards viewing their own culture as preferable, even morally superior to others.

The Negative Effect of Ideologies

I believe that ideological movements in wealthy Western nations are the most serious threats to future progress. In particular, ideological movements within the United States, which has been the engine of progress for the last century-and-a-half, are concerning.

Though radical ideologies seem very different from each other, they reflect the same mentality. Ideologues on both the right and left divide everyone into groups based upon some demographic characteristic. The group might be based upon ethnicity, race, class, gender, religion, sexuality or nationality. Each group is designated as either "good" or "bad." In their world individuals do not exist; only groups. The actual characteristics of individuals within each group are irrelevant.

The "bad" groups typically are successful people, while "good" groups are less successful. Ideologues believe that the bad people achieved their success by doing bad things to the unsuccessful. And the good people have done poorly because of the terrible things that bad people have done to them in the present or recent past.

The goal of the government, in their view, is to punish the successful group while boosting up the unsuccessful group. Progress, to the extent that they believe in it, consists of redistribution, punishment and shame.

As we have seen, this is an extraordinarily inaccurate and dangerous view of human history. Different outcomes between people are largely accounted for by geographical variations in the distant past, which constrained food production. How a society acquired its food, in turn, structured their entire society. This resulted in some societies being able to innovate new technologies, skills and social organizations and others to not be able to. People that were lucky enough to have "good" geography have been successful, while those that had "bad" geography stagnated in poverty for millennia.

Rather than hate the successful, we should feel fortunate that a few societies have been able to break out of those geographical constraints. We should also feel fortunate that Industrial societies have innovated technologies, skills and organizations that enable less successful people to finally overcome their historical constraints.

KEY INSIGHTS

Radical ideologies undermine the foundations of progress by overly centralizing power and stifling diversity of opinion. This undermines the experimentation necessary to promote progress.

For the first time in history, less successful people can go from poverty to progress by copying more successful people in one generation. Less successful people and those who pretend to speak for them should be grateful that finally a way out of millennia of poverty has been created. All they need to do is copy the successful.

Ideologues preach a shortcut for what the right government policies can do to help the less successful. But there are no shortcuts. The only proven strategy for success is to copy what has already worked. Dreaming up better pathways to success will lead us nowhere.

Ideologues on both the left and right claim to speak for certain groups and know what is best for them. In reality, they preach policies and attitudes that doom that group and all others. If we focus all our efforts on resenting the successful and punishing them, we hurt everyone, particularly less successful groups.

While ideologues on both the right and left claim to be able to see the world more clearly than everyone else, their ideas are merely the intellectualization of anger and resentment. This anger and resentment colors everything they see and distorts their entire view of history and the human condition. That anger and resentment harm every individual or group that adopts their views. That anger and resentment are key threats to progress for all of mankind.

The goal of many of these ideologues is to provoke anger and resentment among the less successful, and shame among the more successful.

They believe those powerful emotions will provoke sudden action that creates a better world.

But we already know how to create a better world. Our goal should be to maintain the preconditions that enable the vast, decentralized problem-solving network to solve problems every day. The powerful emotions provoked by ideologues undermine the network because they undermine the ability to cooperate and copy others based upon what works.

While we need positive thinking to keep progress going, ideologies emphasize the negative. There is an enormous amount of evidence that optimists, as long as they are also grounded in reality, are far more successful than pessimists. Pessimism can easily become a self-fulfilling prophecy as it constantly throws up psychological barriers that discourage action.

While future progress needs people to understand their individual role in promoting success, both for themselves and all of society, ideology encourages people to portray themselves as victims held back by other groups. This victim mentality encourages people to ignore what they have accomplished or, more importantly, what they can accomplish regardless of whatever barriers exist.

While progress requires keeping an open mind, a thirst for knowledge and a willingness to copy what works, ideology closes the mind. Ideology creates the false perception that the world is very simple, it is well understood and all problems are subject to quick fixes. Ideologues believe that solutions already exist, and all that is required is the moral conviction to implement them.

While progress requires that institutions experiment with widely differing solutions, ideology encourages a group-think mentality. Ideologues believe that all good ideas come from within their group and that everyone outside that group lacks the necessary intelligence, courage or moral conviction.

Ideological movements also gradually shut down differences of opinion within the group. This creates a self-limiting dynamic, a downward spiral of conformity that ultimately undermines every ideological movement.

While progress requires that different people with different views and skills focus on solving different problems, ideology encourages the belief that we only have to get rid of the people who are interfering with the implementation of the correct policies. Ideologues believe that we already know what works; we just have to have the moral conviction to implement them over the protests of less enlightened people.

While progress requires problem-solvers who experiment with different solutions, ideologues encourage people to believe that the world is divided between those who speak the truth and those who speak lies.

While progress requires that people constantly compare the current situation with the actual alternatives in existence, ideologues compare reality to a pure vision that only exists in their heads. Reality can never compare favorably to a vision that does not exist. In visions, all trade-offs and conflicting interests disappear into the void of perfection.

But when we compare our reality with what has existed in the past, our current situation does not look so bad. In fact, it looks pretty darn good. An understanding of previous progress can be an incentive to keep working on future progress.

While progress requires practical problem-solvers, ideologies encourage grand visions. Practical solutions to make society just a little better are of no interest to ideologues. Only dramatic transformations are acceptable.

While progress requires that we divert resources towards solutions that will make people's lives better, ideologues wish to divert a huge proportion of our resources towards implementing a utopian vision. Ultimately, that vision can never be achieved and the attempt to do so will waste valuable resources that could be used to solve more practical problems.

While progress requires an understanding of history to learn what works and what previous attempts have failed, ideologues have strong amnesia. History is rewritten as a series of noble failures that would have worked if only "bad" people did not sabotage them. Ideologues believe that their next attempt will have better results. And the very real progress that came from more pragmatic experimentation is completely ignored.

While progress requires a wide variety of opinions, ideology encourages people to shut down differing opinions. The assumption is that a diversity of opinions being voiced only slows down the "correct" policies from being implemented. The more logical the argument of those differing opinions, the more important it is to shut those opinions down.

While progress requires small-scale experimentation and constant iteration based on results, ideology encourages people to ignore results. Because ideologues view the world as a moral struggle between the righteous and the evil, results do not matter. In particular, whether a policy works or not is trivial in comparison to the intentions of the people who originally implemented those policies.

While progress requires decentralization of power so that political, economic and religious elites do not warp experimentations for their self-interest, ideologues seek to centralize power. For ideologues, this makes perfect sense because they believe morality dictates very clear action. Decentralization of power only interferes with the ability of the ideologues to implement their grand visions.

Ideologues fundamentally misunderstand the causes of progress. They apparently believe that the world improves when those who are more enlightened than the rest of us take a public stand and attempt to persuade others through moral argument. As the persuaded become the majority, then the government will adopt policies that make the world a better place.

Ideologues promise shortcuts that enable entire peoples to suddenly jump to another type of society. There is no evidence from history that this is possible. Even worse, they undermine what does work: the vast decentralized problem-solving network that is a modern society.

When ideologues first attempt to seize power, they focus on youth groups and education. Their goal is to thoroughly indoctrinate the next generation so that they cannot even conceive of another viewpoint. The Nazis did this via the Hitler Youth. The Communists did this via the Pioneers and Komsomol.

Teaching children that the best means of achieving progress is political activism and loyalty to the state inflicts terrible damage. Rather than learning the skills and values necessary to flourish in modern society, ideologues teach that the successful achieve their success via exploitation and privilege. Their project will inevitably fail to transform society for the better, but it will cause a great deal of suffering before it collapses.

Ideologues cannot see that progress comes from very mundane individual actions. Progress comes from children and young adults learning skills that are valued in the marketplace, by perfecting those skills by working years at a trade, by cooperating with others in an organization, and by moving to geographical regions that offer greater economic opportunities and teaching those skills and values to their children.

Progress also comes from engineers designing products that are just a little bit better than last year's model, by entrepreneurs founding new businesses to innovate new products and more effective business models, by managers who help to scale up fledgling businesses into enterprises capable of competing on the world marketplace, and by skilled workers solving the little problems required in implementing new technologies, skills and social organizations.

All of this happens organically when the right conditions exist. As we have seen in previous chapters, some of these conditions are geographic, some are political, and many are technological.

Government policy plays a role in this, but far less than many imagine. Most of the policies that promote progress are very mundane and do little to excite political activists: investments in vaccinations, preventative medicine, sanitation, transportation, energy, education in basic literacy and numeracy as well as specialized engineering and entrepreneurial skills.

The big social programs that take up huge amounts of funding do relatively little in comparison, except to redistribute the gains of progress. Indeed the funding of these programs is dependent upon the economic development that progress generates.

Ideologues have a strong self-interest in society not believing in the progress that I am discussing in this book. Think about it. If everyone believed that the world is better than ever, it will likely continue to get better in the future and there is relatively little that government can do to improve the situation, why would anyone support radical politics?

Belief in progress is an existential threat to any radical movement, so they must promote a view of modern society that is dark and pessimistic. They know that they must promote a fear of the future, so they can portray themselves as the protectors of society. They realize that pessimism and fear are key to their achieving power.

Critical Theory Movement

While previous threats to progress have come from totalitarian movements such as Communism, Nazism, Authoritarian Socialism and Fascism, a new totalitarian movement is currently gaining rapid ground within the United States. This movement goes under many different names, including Wokeness and Social Justice, but it is best understood as Critical Theory.

As Helen Pluckrose and James Lindsay have documented in their book *Cynical Theories*, critical theorists see a world without individuals. In their mind, only groups exist. They view the world as permeated with Power and Domination, where groups manipulate language to oppress other groups. Critical theorists see all differences in outcome between any groups or individuals as a sign of linguistic oppression that must be ended.

The flavor of Critical Theory that is most popular is Critical Racial Theory, which preaches that whites dominate other races with their manipulation of language. But it is important to understand that this ideology is not restricted to race. It is gradually being applied to every domain of human knowledge and human relationships. Critical theorists have played the rhetoric trick of playing on people's ethnic/religious/racial identities to heighten resentment towards any group that is more successful than average.

Critical theory has been applied to an ever-expanding number of domains: race, gender, sexuality, disability, obesity, science, mathematics and logic. Critical theorists believe that all these domains need to be subjected to relentless criticism (hence the name) until they are hollowed out and they collapse of their own accord.

Then somehow a new society will be reborn without inequalities of any type (they are very vague on this point, for obvious reasons). It would be hard to come up with a theory that is worse for promoting progress, individual success and happiness.

Critical theory is particularly insidious because it does not just play on one group identity. Nazism and Fascism played on ethnic identity. Communism and Socialism played on class identity. Critical theory plays on all identities, and, even worse, it is constantly inventing new identities that never existed in the political domain.

Because Critical theory plays on all identities and creates justifications for why the disadvantaged should resent the successful rather than copying them, it represents a threat to progress. Just as the Communists and Nazis tried to sabotage progress, or at least redirect it towards profoundly illiberal ends, so does Critical Theory. The fact that it is strongest in the United States, a nation that has played an important role in driving progress over the last two centuries, makes it particularly dangerous.

I believe that Critical Theory will fail because it is based upon such obvious nonsense. It cannot possibly survive questioning or rational inquiry. That is exactly why Critical Theory attacks those concepts. People will quickly realize that no one is safe from mob attacks, real or digital, not even those groups who are supposedly being helped.

People will quickly realize that Critical Theory has no solutions; it has only relentless criticism of everything but itself. People will realize that those who embrace the ideology turn themselves into unsuccessful and unhappy people who alienate everyone around them.

Conclusion

For virtually all of history, humans lived under constraints that made progress impossible. Their societies were trapped at a level of development far below what we consider poor today. Most important were geographical constraints that determined what type of food could be acquired in the natural environment.

These constraints determined which society types could evolve in that natural environment, which in turn greatly affected the rate of innovation. Geographical distance also made it difficult for most societies to copy successful innovations in distant societies.

Human societies have also largely unintentionally erected man-made constraints on progress: extractive institutions, cultural distance, ethic/religious/racial identities and radical ideologies that breed resentment against the successful. Both geographical and man-made constraints have declined dramatically over the last few generations, leading to the progress that we are currently experiencing. Despite this fact, we need to be aware of those constraints against progress that still exist and need to be diminished so that all people can enjoy the benefits of our progress.

CONCLUSION

Throughout this book, I have made the argument that progress is the outcome of a long evolutionary process that started with the Big Bang, created life on our planet and enabled the evolution of complex biological organisms. Humans pushed this process further by adding on Cultural evolution. Cultural evolution enabled humans to construct complex societies with large numbers of technologies, skills and organizations, but it did not immediately elevate the standard of living of the masses.

While highly innovative compared to other species, humans of the past have been heavily constrained by geography. Geographical constraints determined how a society could acquire its food. In particular, the biome that a society inhabited and its access to domesticable plants and animals largely determined whether agriculture based upon animal-drawn plows could evolve.

How a society acquired its food, in turn, placed powerful constraints on how rapidly the society could innovate technologies, skills and social organizations and copy the innovation of others. Where geography made animal-driven plows possible, complex Agrarian societies evolved. Agrarian societies drastically increased population sizes, enabling higher rates of innovation.

Where geography made animal-driven plows impossible, humans could not evolve past less complex types of societies. Those less complex societies had no chance of evolving the Five Keys to Progress until the arrival of European settlers.

Even in geographical regions that could support Agrarian societies, powerful political, economic and religious elites constructed institutions that extracted the food surplus to the benefit of themselves rather than trade-based cities. This undermined the rate of innovation and hamstrung the potential for evolving the other keys to progress.

Because of these geographical constraints, most societies of the past were trapped in poverty. There was little an individual could do other than survive and live a life almost identical to previous generations.

In some scattered corners of Europe — Northern Italy, Flanders, Netherlands and Southeast England, a few societies were blessed with geographical and political advantages that enabled them to escape the poverty trap. These regions evolved into Commercial societies based upon trade-based cities. As the populations within these societies became larger, more urbanized and more connected via transportation and communication technologies, they became more innovative and more willing to copy the innovations of others.

As these Commercial societies gradually built up the number of technologies, skills and social organizations over the centuries, they evolved four of the Five Keys to Progress.

Once a few societies in Northwest Europe evolved a critical mass of these keys, the result was a vast decentralized problem-solving network. Rather than being caught in the trap of devoting all their working hours to acquiring food, humans could now work together to solve each other's problems. This created progress — a sustained increase of standard of living for the masses — for the first time.

This progress accelerated rapidly when Great Britain introduced the fifth Key to Progress: the widespread use of fossil fuels. Great Britain learned how to transform the incredibly dense energy captured within fossil fuels so that it was useful in agriculture, transportation, communication and the creation of new materials. The addition of vast amounts of energy broadened the trend of progress from a few cities to entire nations and continents.

While the geographical location of the leading edge of progress has shifted over the years from Northern Italy to Flanders to the Netherlands to England and finally to the United States, progress continued and accelerated. While progress started as a few faint candles in the darkness on one corner of the globe, the candles gradually grew brighter, more numerous and more widespread. Today, those tiny dots of light have evolved into powerful searchlights that shed a bright light on entire regions.

Innovations made by Industrial societies since the Industrial Revolution have made it possible for nations to overcome the tyranny of geographical constraints and extractive institutions. These innovations have made it feasible for almost every society to escape the poverty trap. As nation after nation copied the technologies, skills and social organizations of the most successful nations, they have included their citizens in global progress.

Though most geographical constraints to progress have been greatly reduced, there are other barriers to progress that remain. Traditional ethno/religious/racial identities intensify resentments against successful people and nations, thereby undermining the desire of the poor to copy more successful models. Traditional elites understand that progress creates new sources of wealth that potentially undermine their power. To preserve their power, traditional elites deliberately intensify ethno/religious/racial identities against more successful groups.

Radical ideologues on both the left and right also try to intensify resentments of less successful groups as a means to achieving power. Even worse, they seek to dramatically centralize political and economic power and subordinate that power to ideology. This undermines the decentralized trial-and-error experimentation that is necessary to promote progress.

Rather than copy the successful, traditional elites and ideologues want the people to resent the successful. They foster the belief that different standards of living are caused by the successful hurting the less successful.

This stark zero-sum viewpoint seriously undermines the ability of entire peoples to experience progress. The ability of individuals and peoples to ignore the siren call of ideologies and group identities and be willing to copy the successful from other cultures will largely determine whether they enjoy the benefits of future progress.

In whatever domain one chooses, there is one golden rule: copy the successful. While it is not always easy to know exactly what it is about the successful that needs to be copied, copying is the fastest route to success. Unfortunately, many individuals and even entire peoples waste their time and energy resenting the successful, trying to hurt them or isolating themselves away from the perceived harm. This viewpoint undermines their chances of enjoying the benefits of progress and leading a happy and successful life.

The Future of Progress

So will progress continue in the future? My life-long study of history and public policy has made me skeptical of predictions of the future. Even experts who are very confident in their opinions have a very poor track record of success. Predictions that involve sudden changes in past trends are particularly prone to error.

A big part of the problem with expert predictions of the future is that they usually rely on one causal factor. Experts have a great deal of knowledge, but they tend to focus excessively on "one big idea" and overestimate its importance relative to other factors. By ignoring the vast majority of causal factors that impact human history, they warp the accuracy of their predictions.

In short, the world is so complex that even the most intelligent human (or computer) cannot understand all its complexity. Without understanding its complexity, predictions about the future should be made with a profound sense of humility.

I do believe, however, that it is possible to look at certain trends of the past and assume that they will continue with some level of certainty. This book has described trends and processes that have occurred for hundreds, thousands, and in some cases, billions of years.

Given the enormous length of time that they have endured, it seems very unlikely that they will suddenly change within our lifetime.

Barring some unpredictable events that threaten the extinction of mankind, I do expect the following trends to continue.

There is every reason to believe that our current progress will continue in the future. All of the factors that lead to innovation in technologies, skills and social organization are present and becoming increasingly common:

- The technological base is vast and constantly increasing in size.
- The population of humans on planet Earth is larger than ever.
- The proportion of humans living in cities is at unprecedented levels.
- The diversity of skills of people living in those cities is also extremely high and rising.
- The level of education, particularly in engineering, is also rapidly increasing.
- Forced labor has been all but abolished.
- Agricultural production per unit of land keeps improving faster than the growth of population.
- Humans are far more connected by transportation and communication technologies than ever before.
- Energy technologies have made fossil fuels affordable for most people.
- Social organizations, particularly corporations, are becoming increasingly numerous, specialized and diversified.
- Industrial production techniques are making products cheaper and more widely available, making it possible for the poor to purchase what only the rich could just a few years ago.

While progress may be interrupted in individual nations or regions, the wide dispersion of all of these factors across the globe makes it unlikely to affect the overall trend towards progress.

Most potential negative events or trends — wars, famines, epidemics, climate change, recessions, coups, revolutions, or genocide — are unlikely to affect progress in the long run. They may seriously disrupt progress in one nation or even an entire region, but at most, the locus of progress will shift to another region.

Modern human societies are simply too adaptable. The global recession in 2008 is an example of a large negative event that caused an intense downturn, but then humans adapted and progress continued. Within less than a decade, it was as if the recession had never happened to the vast majority of people.

Many negative shocks will occur, but the progress trend will continue. The world population is too large, too urbanized, too educated and too dispersed around the globe for humans to not be able to adapt. More importantly, the technological base is too large, our skill sets are too broad, our organizations are too numerous and the rate of innovation is too rapid not to have a long-term impact.

Some extinction events could bring progress to a permanent halt because they have the potential to annihilate cities, agriculture and our ability to use fossil fuels — asteroid impacts, gamma-ray bursts, or nuclear war — but these events seem relatively unlikely in the near future. In short, nothing less than the annihilation of 95+ percent of humanity will interrupt progress.

Nations that are least developed today will likely experience the greatest progress in the coming generations. They will do so largely by copying technologies, skills and social organizations from wealthier nations and then adapting them to their local environment.

When viewed across the globe, this trend will appear long and slow, but individual nations will probably undergo significant changes quickly. Just as Sweden, South Korea, Taiwan, China, Botswana, and Chile more than doubled their standard of living within one generation, so will other countries in the future.

Some nations will experience little or perhaps no progress, largely due to ideological regimes that promote ethnic, religious or racial resentments internally or against the United States and Europe. Some, like Syria, may even implode into a civil war that sets them back for generations.

Because political regimes can change rapidly, it is hard to predict which nations will fall into this category. North Korea and Venezuela are two examples of political regimes that have recently devastated their countries. It seems very likely that a few nations will follow this path, but I am hopeful that they will be outliers.

Inequalities of income, wealth and political influence will continue within all nations. Disparities in income, wealth and military power between nations will also continue. Because those inequalities will align with ethnic, racial and religious divides, these disparities will continue to be a source of divisive ideological conflict and political instability. The ability of political leaders to promote copying the successful rather than promoting resentment of the successful will be one of our central challenges for generations to come.

> **KEY INSIGHTS**
>
> Barring an extinction event, progress is very likely to continue in the future. But it is up to you to decide whether you want to participate in it and enjoy the benefits of doing so. What do you choose?

Believing In Progress Works

So far in this book, I have tried to appeal to your intellect by presenting facts and making logical arguments for the existence of progress. As I wrap up this book, though, I want to appeal to the more emotional parts of your brain. I want to make the case that even if the cynics of progress are largely correct, that *it is still better to think and act as if my argument is true.*

This book is not a self-help book. It is a book about history, technology and social science. But as I wrote it, I came to see how these concepts can help people to live more successful and happier lives.

It is all too common for people to sabotage their own lives because of cognitive distortions that affect how they view the world. Those cognitive distortions affect the choices that a person makes. Those choices, in turn, have an impact on their future success and happiness.

If people believe that we now live in the best time ever and things will likely continue to get better, it helps to clear away the negative thoughts that hold people back. By seeing life as good, you can make choices to improve your own life. I believe each of us can lead better lives.

Even if my argument is only half true, and the cynics are half correct, which viewpoint enables you to have a better life? If you believe that conditions are bad, they are getting worse and there is not much you can do about it, what is the chance that you will have a successful and happy life? What tools does that viewpoint give you? All you can do is complain endlessly about how bad things are and do nothing to improve your life.

But if you believe that there is at least some progress and there is something that you can do to take part in that progress, you have laid the foundations for a better life.

I guess that the bulk of the readers of this book will be educated professionals from affluent nations. I encourage you to look at your life and see how many of the decisions you made in your past enabled you to be where you are today. Whether you were conscious of it or not, you took actions to improve your life as if success and happiness were within your grasp and there was something good going on in society that you wanted to be part of.

A progress-based perspective focuses on what previous generations have achieved and what you as an individual can do to be a part of it. We need to combine the passion to make the world a better place with results-based experimentation at scale using methods that worked well in the past.

Even if the cynics are correct and progress does not exist, what is it in your best interest to believe? What is in your children's best interest to believe? What is in your best interest for other people to believe?

Understanding the extraordinary difficulties that previous generations lived under sets our current problems into context and makes them appear less daunting. For me, seeing that previous generations made enormous sacrifices for their children and other members of the following generations generates a profound sense of gratitude. That sense of gratitude gives me the motivation to do the same for future generations.

Feeling that we humans are part of a common mission to improve our lives can give each of us a sense of identity. Realizing the enormous progress that we are living in gives one hope for the future.

Understanding that there are simple steps to improve one's own lives, gives each of us a sense of agency.

If we have an awareness of the progress that previous generations have passed down to us, a feeling of gratitude for benefitting from their efforts and a willingness to learn how they achieved that progress, we are in a much better position to solve the problems of today.

I am not a religious man, but I must admit that viewing the entire scope of history gives me a profound sense of awe. We evolved out of simple sub-atomic particles without any consciousness or thought to become humans capable of constructing extraordinarily complex societies. Those societies eventually created widespread progress for humanity.

As a natural part of every day, those societies constantly solve problems to make other people's lives better. And we are all lucky enough to be born at a time when the collective achievements of previous generations have made our lives better than it people's lives have ever been.

Understanding previous and current progress also gives me great hope for the future of mankind. We have solved so many thousands of problems in the past, that it gives me confidence that we will solve current problems. The challenges that we face today are not crises. They are, in fact, trivial obstacles in comparison to the sheer struggle for survival that dominated most of human history.

Progress is already a solved problem. We do not need radical changes. We just need to copy what works. We just need to "keep on keeping on."

Previous generations have already done all the heavy lifting by building societies that naturally promote progress. They have passed onto us all the winning lottery tickets that they accumulated over the course of their lifetimes. We just need to cash them in.

Thanks for taking the time to read this book. I plan to follow up this book with a series of books about how progress evolved and accelerated over the last millennium. I hope that all the books in this series convince you that we live in a world of great progress, that progress has a positive impact on people's lives, and that we can all make choices that increase the probability that we will benefit from that progress.

Website for Book Series

You can find additional content related to this book series at:
frompovertytoprogress.com

With a **free** subscription to this website, you get:

- Large discounts on audiobooks and e-books
- Free book samples (E-book and audiobook)
- Access to Podcasts and Blog posts about related content
- Plus more.

Additional Content and Graphics

To keep the price of this book as affordable as possible, some content has been moved to the author's website. On this site, you can find:

- Full-color graphics
- Extra content and graphics
- Bibliography
- Recommended Reading List
- Access to summaries of related books
- Books that Influenced Author the Most

To access this content, scan the following QR code or go to **frompovertytoprogress.com** and search for "Additional Content"

Credits

Editor: Hugh Barker (can be contacted via Reedsy.com)
Cover Designer: David Provolo (can be contacted via Reedsy.com)
Book Layout: David Provolo
Audio Technician: Paul Moody (can be contacted via Upwork.com)
Audiobook Narrator: Michael Magoon

Special thanks to Tyler Cowen and Emergent Ventures for providing funding for the publication costs of this book series.

Except as listed below, all of the graphics were created by the author using source data listed on the graphic.

The following graphics were used under Creative Commons 1.0:
Cover photo taken by Mohamed Hassan and downloaded from pxhere.com.

The following graphics were used and modified under Creative Commons 3.0:
UN Human Development colored line graph (http://hdr.undp.org/en/content/copyright-and-terms-use)

Biome map: Data from World Wildlife Fund

Map from: Columbia University Center for International Earth Science Information

NASA Socioeconomic Data and Applications Center

The following graphics were used and modified under Creative Commons 4.0:
Absolute Poverty (using data from gapminder.org)

All line graphs and other graphics with the "Our World in Data" logo on top right were created and downloaded from https://ourworldin-data.org/

The following graphics were used by permission from the original creator:

Growing Seasons map (http://geoagro.icarda.org/)

ABOUT THE AUTHOR

Michael Magoon received a BA in History from University of California at Berkeley and a PhD in Political Science & Public Policy from Brown University. He taught university courses for the following three years. Feeling restless with the staid academic life, Magoon decided to make a career change and entered the rapidly growing field of digital technology.

Starting as a Technical Writer and then as a User Experience Designer, Magoon has worked for some of the biggest and most innovative technology companies in the world, including Microsoft, Apple, Intel, Oracle and Verizon. While working in the technology field, he continued reading books about a wide variety of topics including science, geography, technology, culture and history.

Magoon's background in both academia and technology corporations has given him a unique viewpoint on progress and how it affects our lives.

www.ingramcontent.com/pod-product-compliance
Lightning Source LLC
Chambersburg PA
CBHW061133120626
46546CB00005B/1769